UTILIZATION OF FISH WASTE

UTILIZATION OF FISH WASTE

Editors
Raúl Pérez Gálvez
University of Grenada
Department of Chemical Engineering
Grenada, Spain

Jean-Pascal Bergé
Laboratory of Science and Technology of Marine Biomass
Loire-Atlantique, France

CRC Press
Taylor & Francis Group
Boca Raton London New York

CRC Press is an imprint of the
Taylor & Francis Group, an **informa** business
A SCIENCE PUBLISHERS BOOK

CRC Press
Taylor & Francis Group
6000 Broken Sound Parkway NW, Suite 300
Boca Raton, FL 33487-2742

First issued in hardback 2019

International Standard Book Number: 978-1-4665-8579-9 (hbk)

Visit the Taylor & Francis Web site at
http://www.taylorandfrancis.com

CRC Press Web site at
http://www.crcpress.com

Preface

Modernization and increasing capacity of fishing vessels, are ensuring that most fish species are being harvested to their limit. International organizations, such as UN or FAO have warned about the depletion of fish stocks, and have highlighted the need to promote new regulations and management strategies to ensure sustainability of extractive fishing. In light of this, the transformation of fish waste or by-products from into added valued products represents an underexplored opportunity for fish processing industry. This book analyses the problems derived from the management of these materials and reviews the current solutions to up-grade them.

Chapter 1 reviews the current problems relating to the management of fish waste and by-products, focusing on the French West Atlantic coast. So far, fish by-products have been disposed off or reduced to fish meal or fish oil, which do not increase the added value of the final product. Given the nutritional value, new strategies are required to provide a better utilization of marine by-products. Chapter 2 summarizes current methods of extraction and isolation of relevant components, such as restructuring fishas lipids, proteins, enzymes, peptides, chitin, and chitosan.

Among the final products derived from marine matrices, fish protein hydrolysates have food and nutraceutical applications. Their usage in foodstuffs is based on properties, such as solubility, emulsification or sensory properties, which are described in chapter 3. In addition, this chapter discusses the possible correlation between peptide structure and its functionality in the food system.

The development of new extraction technologies and research has led to the identification and isolation of an increasing number of bioactive compounds which are of interest for nutraceutical and pharmaceutical applications. These compounds include peptides, oligosaccharides, enzimes, fatty acids or vitamins and exert a number of biological activities. Chapters 4, 5 and 6 review the current knowledge on marine-derived compounds, mainly peptides, displaying antihipertensive, antimicrobial and secretatologue activities, respectively.

Besides the protein fraction, lipid content of fish varies from 2–30%, depending on the species, diet, season and environment, and geographic variations. Two polyunsaturated fatty acids, eicosanpentaenoic acid (EPA) and docosahexaenoic acid (DHA) are predominantly found in many marine animals, and have been reported to have benefits on human health. Among the benefits of these lipids and other belonging to the omega-3 family, the best known are prevention of cardiovascular diseases and the early human development. Chapter 7 presents current data on composition and seasonal variations of the lipid profile of some commercial species. Furthermore, it explores several methods for the valorization of fish by-products, such as omega-3 fatty concentration or production of structured lipids.

Finally, the last chapter has been devoted to the study of up-grading of fish materials as energy source. Vegetables oils are the most common feedstock in the production of biodiesel, a renewable and environmentally friendly fuel. However, the usage of the agricultural resources for fuel instead of for food, presents both an ethical and an environmental problem. An alternative feedstock is fish oil, which is produced in large quantity by fish-processing industry and has similar calorific value to petroleum distillates.

This book represents the combined efforts of over twenty contributors, who belong to academia and research community. The editors want to express their gratitude to all of them for their valuable expertise and contributions to this book.

Raúl Pérez Gálvez
Jean-Pascal Bergé

Contents

1

By-products from Fish Processing: Focus on French Industry

Anaïs Penven,[1,*] *Raúl Pérez Gálvez*[2] and *Jean-Pascal Bergé*[1]

ABSTRACT

Beyond the technical and technological advances in the conversion of marine by-products into useful products, the economic, social, political and environmental parameters, are to be taken into account to understand why and how bioconversion technologies of fish by-products are applicable. The fish sector is facing several issues such as wild fish stocks reduction while in the same time the cultivate one rises, augmentation of the importations on a global market, consumers behaviors. All those influent parameters affect and complicate the definition of an upgrading strategy for the by-products as they directly depend to the raw material processed.

Thanks to some projects addressed this thematic in recent years, several quantitative and qualitative data about by-products availabilities are

[1] Laboratoire Sciences et Technique de la Biomasse Marine, IFREMER, Rue de l'ile d'Yeu, BP 21105, 44311 Nantes Cedex 3, France.
Emails: anais.penven@ifremer.fr; jean.pascal.berge@ifremer.fr

[2] Departamento de Ingeniería Química, Universidad de Granada, 18071 Campus Fuentenueva, Granada, Spain.
Email: rperezga@ugr.es

* Corresponding author

available. It is time now to propose upgrading strategies and to study their technical and economical feasibilities on different territories. In France, nowadays only two treatment plants are still active for fish by-products processing and they are both located in the northern part of the territory. The comparison of their operation mode allows to identify factors of success or sticking points such as volumes treated, number and proximity of fish processors, qualitative parameters or management. It appears that the optimisation of fish by-products upgrading may depends of many of these factors it is so essential to identify and to study them.

INTRODUCTION

There have been numerous biotechnological advances in the conversion of marine by-products into useful products. In order to give maximum elements of understanding, it is essential to define the framework of this research to understand why and how bioconversion technologies are applicable. It is essential to look beyond the technical and technological advances in the subject and also to take into account the economic, social, political and environmental parameters, which govern all forms of approaches for fish by-products upgrading.

All concerns given to marine resources are relatively new. It began in the mid-90s with the publication of two major reports of the FAO in 1994 and 1995, both dealing with the need for responsible fisheries and minimising wastage, particularly in relation to by-catches. They are still a concern for scientists with regard to environmental issues. The European Commission works on this thematics but the legislation on it is still not well defined. Recently it was proposed to keep all by-catches on board, but this law is still under discussion.[1] To ensure the feasibility of such a regulation, Raul Perez Galvez worked on a prototype of compaction so that these by-catches take minimal space in the fishermen's vessels (Perez-Galvez 2010). As detailed studies on by-catch and discards have already been carried out (FAO 2004), this problematic will not be considered in this paper. Here, we will highlight the various issues concerning the by-products generated by fish processing companies on-land, focusing on French territory.

Problem Targeted

Management and processing of by-products and waste is an environmental, social and political priority for many countries and is more problematic because of the rising production volumes. In France, the food industry alone

[1]Proposed reform of the Common Fisheries Policy, European Commission, 13th July 2011.

generates more than 48 million tons of by-products and waste annually from two main types of transformation: the animal industry and the plant sector. By-products from fish processing are part of the former and account for at least 215,000 tons (0.4% of the entire deposit) according to the estimates for the period 2004–2005 (Perez-Galvez 2009). The volumes concerned may seem low but as it will be discussed in the subsequent chapters, fish by-products contain many compounds of interest in various sectors, providing them a great value which is certainly under-utilized at present. Although many efforts have been made by the food industry to manage its waste, improvements are needed for an effective and specific treatment.

The fish industry is made up of many small businesses scattered in space which creates logistical problems and difficulties in the setting of new waste recovery units. Moreover, a significant amount of waste is generated by supermarkets and hypermarkets. This major sector in France generates large volumes of food by-products resulting from unsold, obsolete or defective products including fish by-products which represent 6% of the deposit (ADEME 2010). Nowadays, by-products from supermarkets are currently directed to incineration causing significant impacts on the environment, which could certainly be optimized.

Difficulties Faced by the Fish Sector

The fishing industry is facing major environmental, economic and social issues. The volume of landings have been declining for several years due to depletion of fish stocks. Production costs are high and rising prices of diesel is part of a trend in the long run even if efforts are made to support the occupations most affected (OECD 2012). To address the overexploitation of fisheries resources, management tools have been developed as part of the Common Fisheries Policy for the European Union, but even so, the level of degraded stocks remains higher than the level of regenerated stocks. According to an estímate by FAO, in 2005, 23% of global stocks were moderately exploited or underexploited, 52% experienced a level close to the maximum sustainable exploitation and 25% were overexploited, depleted or renewed (17%, 7% and 1%) (FAO 2006). Additional measures, including incentives to improve the selectivity of fishing gear, the obligation to change fishing locations, the reduction of by-catch and the gradual reduction of discards in European fisheries are gradually being put in place by the European Commission (Communication from the European Parliament Commission 2007).

The problems encountered by capture activity involve a consideration about the improvements that can be made for processing activities in order to limit impacts on professionals. Management of by-products represents a cost that may be important, it is therefore necessary to study the feasibility of by-products upgradation in not only an environmental but also an economic perspective.

Factors Contributing to Change

In addition to the depletion of fish stocks—and thus landing volumes of them at auctions which gives rise to regular changes in legislation—the seafood industry has evolved in recent years in very many ways due to several factors that have had an impact on raw material processing and thereby on the volume of by-products. These include:

- The development and changes in distribution and market, and
- The changing environment with regards to imports and aquaculture.

These changes particularly affect those who catch (fishing) and the first market (fishmongers) (FranceAgriMer 2009).

Distribution and market

The fish and shellfish level of consumption in France is about 36 kg live weight equivalent annually per person (24 kg of fish and 12 kg of shellfish, crustaceans and molluscs), all types of presentation combined (fresh, frozen, canned or prepared foods) (OFIMER 2010). The main species consumed are: tuna, salmon, hake, cod, sardine, trout, mussel, oyster, scallop, shrimp and Jacques. They now occupy 65% of the seafood market in France. Many farmed fish are among the ten most valuable species: salmon (France is the largest consumer of Norwegian salmon), cod, trout and tropical fish. But farmed fish accounts for only 11.6% of the fish consumed in France, against 77% in China (FAO 2006).

The consumption of seafood has gone up due to meat products consumption steadily declining since the 90s. The various crises in the meat industry, including the largest one—Bovine Spongiform Encephalopathy (BSE)—and the dietary and nutritional concerns, which are increasing over the years, have encouraged the consumption of seafood (COGEPECHE 2009).

However, it should be noted that today culinary practices are evolving. Fresh, frozen, canned or chilled deli products are no longer equal in front of the consumer's eyes. Customers tend to favor the purchase of frozen foods and to a greater extent, frozen deli products (Cayeux 2007). 52% of fresh seafood is sold as fillets, steaks, and cobbles (FranceAgriMer 2012). Distribution must adapt to these new market demands.

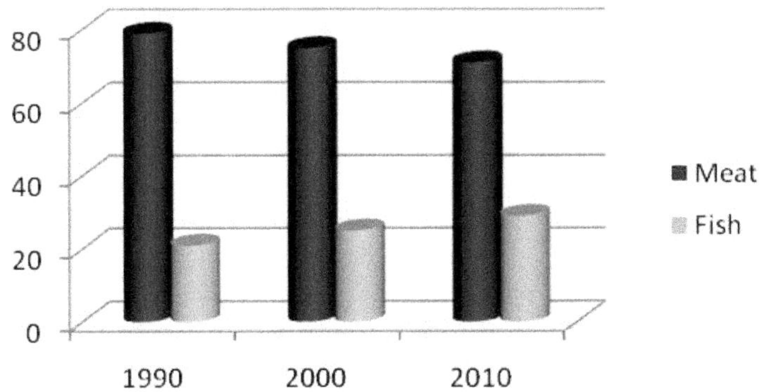

Fig. 1 French households consumption of meat and seafood products in %, FranceAgriMer (2012).

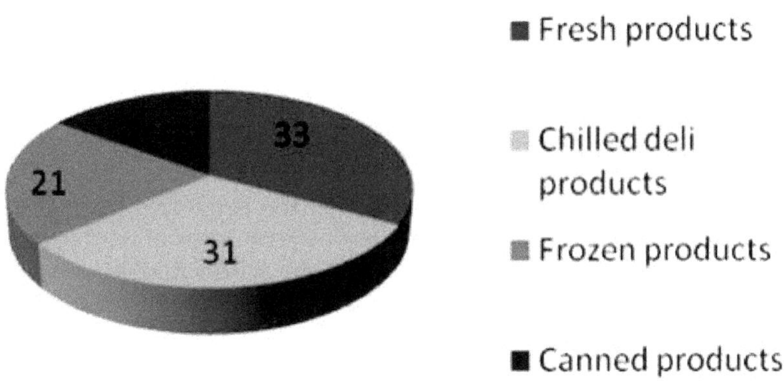

Fig. 2 Households seafood spending in 2011, in %, FranceAgriMer (2012).

This is particularly visible in the supermarkets and hypermarkets. The departments are increasingly being managed in fresh self-service (FLS) in the form of modified atmosphere individual packaging. This induces a decline in the variety of species available for sale (FranceAgriMer 2009).

The above factors show that the management of by-products is a growing issue because if fish is bought before the consumers increasingly transformed, the volume of by-products will obviously in consequence be more. In addition, consumer preferences are tending to transform the very nature of these by-products by limiting the types of species being processed.

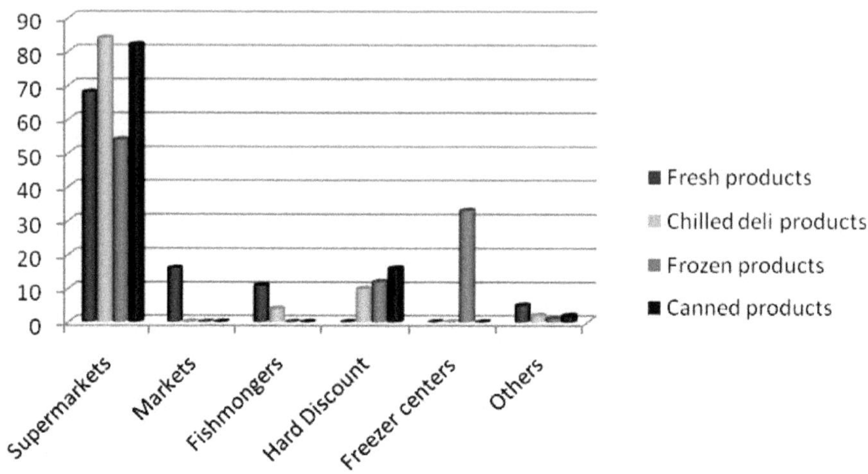

Fig. 3 Purchases of seafood products by French households for home consumption in 2011 (%), FranceAgriMer (2012).

Increase in aquaculture production

Experts have mixed views on the future of the extractive fishing, but no significant increase is expected due to the current status of wild fish stocks. The future supply of aquatic products in the world market rests on the development of aquaculture (FAO 2010).

According to FAO, the aquaculture production must double by 2030 (given the increasing population of the planet) to meet the consumption of aquatic products increasing from 12 kg to 17 kg/year per capita between 2006 and 2008. Already, breeding aquatic products represent almost 46% of the current world fish market against only 9% in 1980. The increase in production is largely due to Asia, China covering more than 70% of global aquaculture production.

The majority of the cultived species are freshwater fish, which is normally raised in ponds and shellfish farmed in coastal areas. On-land fish farms raise species with short food chains (carp, tilapia) whose production tends to increase. Significant species which have been successfully cultivated are shrimp in brackish water in the tropics (Ecuador to Thailand), catfish in cages or ponds (pangasius in Vietnam), salmons (Atlantic salmon and rainbow trout in seawater of Norway and Chile) and marine species in the Mediterranean Sea (bass, sea bream). In summary, aquaculture production is now viable for about thirty marine species.

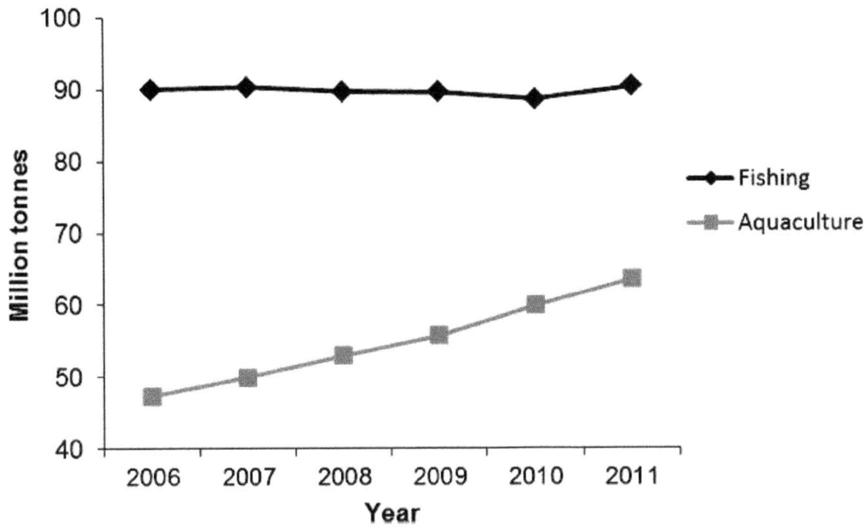

Fig. 4 Fishing and aquaculture evolution in million tons, FAO 2007.

In France, aquaculture consists mainly of trout farming in freshwater with an annual production of 36,611 tons produced by 500 farms, each producing less than 200 tons/year. The marine fish farms sector in France is mainly constituted of bass and sea bream with 4,817 tons produced annually (of which 60 million fingerlings are sold mainly for export). It has not been able to develop due to competition from other countries including Greece and more recently Turkey.

Intensive production is based on the use of blended food, mainly from fish meal and fish oil from industrial fisheries. The growth of aquaculture creates tremendous pressure on the fishing market whose supply is limited and fluctuating as it depends on weather. Production of fish meal from by-products is therefore an alternative that makes sense. Nowadays, by-products represent only 25% of raw materials used to make fish meal and fish oil, but this proportion is increasing for several years (IFFO 2009).

The role of importations in the French fish market

In the context of the stagnating contribution of fishing and national aquaculture, recourse to imports to supply the French market has continued to grow. French importation of seafood represented 1.1 million tons in 2011 (Ofimer 2012). Domestic production covers only 45% of consumption. Imports amounted to 3.6 billion euros in 2005 and 4.8 billion euros in 2011 and came from Europeans countries (including Norway) and the rest of the world at equivalent parts.

The importation of fish fillets, which are mostly raw materials for the processing industry, accounts for 44% of the deficit value. The remaining deficit is distributed almost equally between whole fresh fish and processed fish. In terms of species, the primary deficit is attributed to salmon. Also to be noted is the rapid growth of imports of fillets of freshwater fish: Nile perch from Uganda and Tanzania and pangasius from Vietnam.

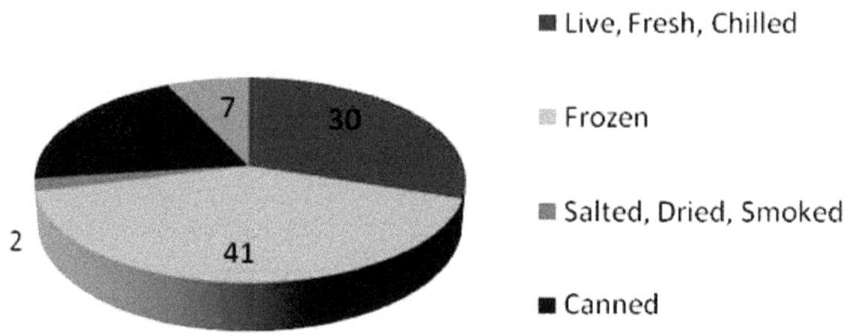

Fig. 5 Imports by presentation in million tons, FranceAgriMer 2012.

REPORT ON FISH BY-PRODUCTS FROM FRENCH PROCESSORS

Terminology, Definition of Waste, by-product and Co-product

First of all it is essential to identify clearly the biomass concerned. The distinction between a waste, a by-product, and a co-product is sometimes thin and definitions are many.

The simplest is to remove the concept of "waste", as waste by definition cannot be upgraded. Since the International Convention of Bâle, in EU, waste is considered as 'any substance or object the holder discards, intends to discard or is required to discard' [European Directive (WFD) 2006/12/EC]. This definition is negative and does not imply any value addition to the products resulting from the processing of raw materials. However, "waste" is still widely used in common parlance to define by- or co-product.

Difference is not always made between by-product and co-product, although in some countries the law provides two distinct meanings.

The term "co-product" means products derived from fish produced along with the final product ultimately intended for human consumption. We can cite, for example, monk-fish cheeks or some fish liver; which can be consumed with or without subsequent processing. These co-products are however then considered as food and therefore are subject to the same regulations.

Fish by-products represent the remainder of the products resulting from the processing of the raw material that will not be valued as human food. For the French the veterinary services department, by application of Regulation (CE) N° 1774/2002, that came into force on 1 May 2003 in the context of strengthening food security, the term "co-product" has no statutory existence. The set of materials derived from the production of finished products are considered as by-products.

The by-products are then separated into three categories based on their potential risk to human health, animal health and the environment:

1. The first category (C1) is composed of by-products that present a significant risk to public health. These must be collected, transported and identified without delay and are destroyed by incineration or landfilling after processing and marking.
2. The second category (C2) includes by-products presenting a lesser risk to public health (products containing residues of veterinary drugs, for example). These products are disposed of by incineration or landfilling after processing or can be recycled for uses other than animal feed (organic fertilizers, biogas, compost, etc.).
3. The third category (C3) incorporates animal by-products which present no health risk and includes parts of slaughtered animals fit for human consumption. Only Category 3 material can be used in animal feed, after application of appropriate treatment in approved processing plants.

By-products of any value are those from category 3. We shall therefore talk only of category 3 by-products, except for application in human food where it is more appropriate to use the term "co-product".

Generation of by-products by the Fish Processing Industry

Processors of seafood are working to add value to raw materials by putting them in line with market expectations. Therefore, seafood can undergo numerous processing steps such as, filleting, heading, gutting, skinning or cutting. Fish by-products are composed of heads, viscera, trimmings, fishbones or cartilage, hides, tails, and eggs, whose average proportions are described in Fig. 6.

Three major types of industries generated by-products include:

1. *Fish trade.* While generally all species landed can be transformed by the wholesalers, only white and cartilaginous fishes are usually processed at this scale.

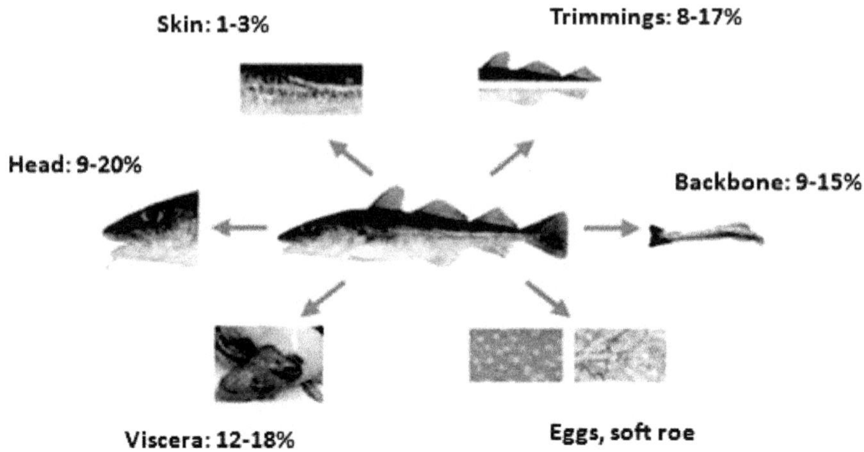

Fig. 6 Average proportion of fish by-products, Dumay (2006).

2. *Canning industry*. The canning industry uses pelagic species as their fatty flesh is ideal for conservation. The species used by canners are mainly sardines, mackerel, herring and tuna.
3. *Smoking industry*. Although the fish smoking industry is diversifying more and more using many species, it is the salmonids (salmon and trout) that are predominantly used by these processors.

Processing practices depend on several factors such as species, market destination, proximity of processors, historical territorial habits, etc. As an example, in the southern French Atlantic coast, species landed are usually sold without any processing. Red mullet or sole which are specific species landed in this area are traditionally sold as whole fish.

Methodology Employed to Estimate French Fish by-Products

Recent studies on the evaluation of volumes of marine by-products have developed estimation techniques. An indirect method that uses conversion coefficient will be first described and then a direct one dealing with surveys of professionals will be exposed.

Indirect method. In 2004, OFIMER published the results of a study conducted in a period of 6 months that allowed to present the first figures of French national assessment of marine by-products volumes in the territory (Andrieux 2004). This study also helped to provide an inventory of existing upgrading methods of marine by-products.

The method developed for this study was however, complicated as too many parameters were taken into account. It was based on available statistics from OFIMER as auction sales in fish markets and imports-exports of each species. Because in France these data are collected and centralized by OFIMER, it was possible to develop such a method.

The estimates were made according to the following procedures:

1. Fourteen representative species were studied as they are usually processed by three kinds of industries: the fish trade (fishmongers), the canning industry and the smoking industry. The first point was to identify those species, thanks to OFIMER's statistics.
2. The second point of this study was to determine which part of the volumes of fish available on the territory was available to be processed. Thus, the trade balance (whole fish only considered) was studied as shown below:

   Apparent Consumption = Fish landings + Imports – Exports

3. Previous studies on fish consumption in French households have shown proportion of each species that are generally bought processed. Thus, the processing rate for each species has been evaluated. However, all the species available are not transformed by the industries as whole fish are still sold.
4. Conversion coefficients then were applied to each raw material to estimate the amount of by-products generated by each processing industry. Those average coefficients are well known thanks to the available bibliography.

Taking into account all those parameters, 150.000 tons of by-products have been estimated on the French territory. An evaluation extended to 20 species has identified 215.000 tons of by-products (Perez-Galvez 2009).

Direct method. In 2008, the "Gestion Durable" project started on the same topic even if the goal was different. Here, the main objective was to demonstrate that a better management of fish by-products could help in better governance of the fishing ports. This project was designed primarily to re-evaluate the volumes of by-products to precise the data previously acquired.

The precise location of the deposits appeared here as a key element because the goal was to understand the structure of the marine by-products industry from several angles: structural, political, economic, geographic and technical.

Given the estimation method proposed by OFIMER in 2004, the Gestion Durable project chose to proceed another way using surveys with professionals to clarify the data regarding some factors that were only estimated in previous works, such as purchase/imports and processing rates.The goal was to clearly identify if the apparent processing of fish raw material corresponds to the estimation previously done thanks to OFIMER's work.

1. A census of all the seafood industries was conducted for four regions of western France representing 47.5% of the landings leading to the identification of 393 industries.
2. Enquiries have been elaborated to obtain quantitative data for all those industries. They have been designed to cover the maximum points regarding all the subjects involved in the project. The objective of these enquiries was to quantify and qualify the raw material effectively processed, the type of processing done and the by-products resulting from the processing.
3. Furthermore qualitative data had been collected on sorting, storage, packaging and preservation. That is part of the innovative aspect of the project.
4. The last point of the survey aimed to collect information on the procedures for collecting by-products to identify costs, proximity to processing sites, frequency of collection, and problems encountered.

Results

The first estimations given by OFIMER in 2004 reported 144,407 tons of by-products on the French territory. The Gestion Durable project published its results in 2011 for the four regions of West Atlantic coast (Basse-Normandie, Britanny,[2] and Pays de la Loire, Poitou-Charentes). The results are based on 167 companies enquired (67% fish trade, 33% smoking, canning and other industries) representing 42% of the companies (167/393). 45,138 tons of by-products were clearly identified during the project. These deposits were declared by processors without making use of any conversion coefficient.

The quantitative results of these two methods are difficult to compare because the division of regions is not the same, so the statistics are dissimilar. For example, if we consider the data including the areas addressed in the project "Gestion Durable", the volumes identified by OFIMER reach 129,990 tons of by-products because they take into account the Nordic and the Aquitaine region. It is also difficult to compare are the two have two completely different methodologies, and they have also been carried out five years apart.

However, it is still possible to discuss similar results and some of the divergent points:

[2]To facilitate the enquire, Bretagne—a region where fishing activities is important place—has been divided into three areas corresponding to the north, west and south coasts. The results obtained for Bretagne correspond to the addition of those three areas.

- As a preliminary result, it has to be noted that the most significant response rate were obtained in the survey areas further north during the enquiry phase of the "Gestion Durable" project. As an example, the enquiry was conducted on 86% of the industries in Basse-Normandie but only 12% in Poitou-Charentes. Some factors may explain this like lower landings, species landed, processors or territorial habits. This can be correlated with the volumes identified by both methods. Indeed, for the study conducted by OFIMER, the largest volumes of by-products are located in northern France, the regions of Bretagne, Normandie and Nord accumulating 80% of volumes, that is a logical correlation given the French landings.

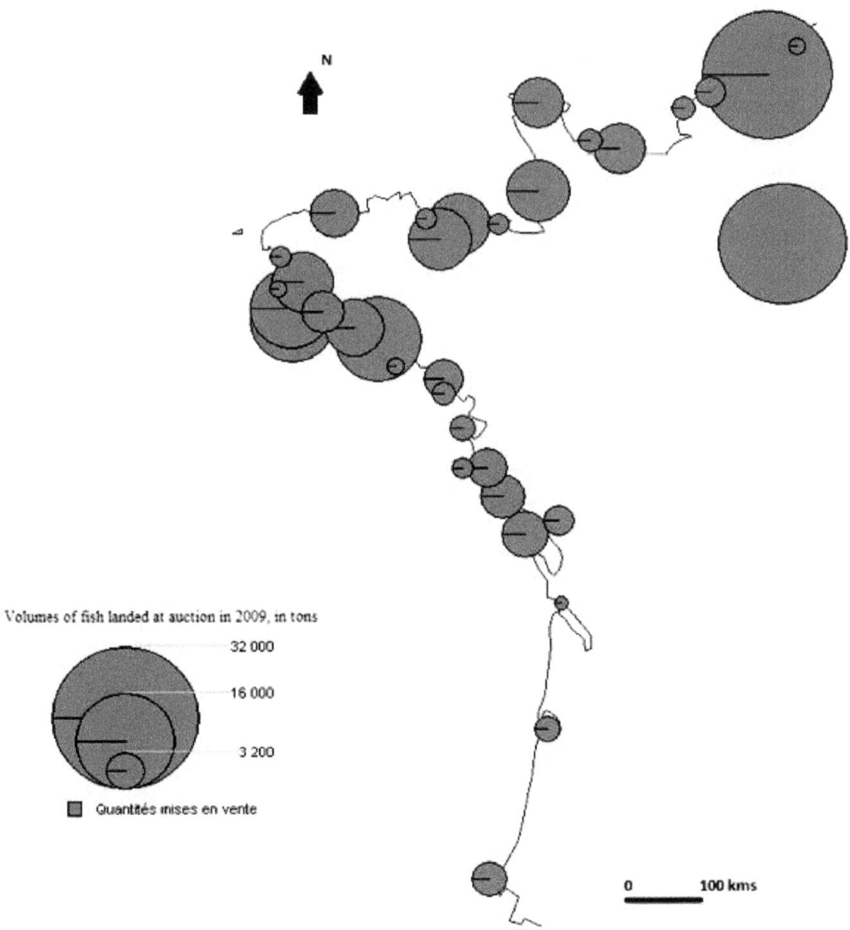

Fig. 7 Sales at auction in 2009, in tons OFIMER (2010).

The results obtained during project "Gestion Durable" tends to confirm this trend as 85% of the deposits were identified in Brittany, which represented 65% of the companies enquired. Half of the total volumes was concentrated in western Brittany.

- By-products represented in 2004 were 46% of the landings in the OFIMER's study, while they represent only 32% of the landings according to the survey results of the Gestion Durable project in 2009.
- For OFIMER, fish trade is the most important generator of by-products, comprising 70% of the total volume. The results of the Gestion Durable project are not only more nuanced but it to place fish trade at 58%. The canning and smoking industry then represent respectively 27% and 15% of the identified deposits.
- The OFIMER's study was able to clearly identify the type of by-products generated. White fish by-products accounted for 40% of the volumes followed by salmonids with 31%, pelagic fish 15%, cartilaginous 7.5%, and other white fish species 6.5%. For the Gestion Durable project, the type of by-products could not be identified as 37% of the by-products deposits which were classified as "various". For by-products that were identified, 28% came from pelagic species, 15% from white fish, 15% from salmonids, and 2% from cartilaginous. The remaining 3% were from other types of species such as cephalopods.

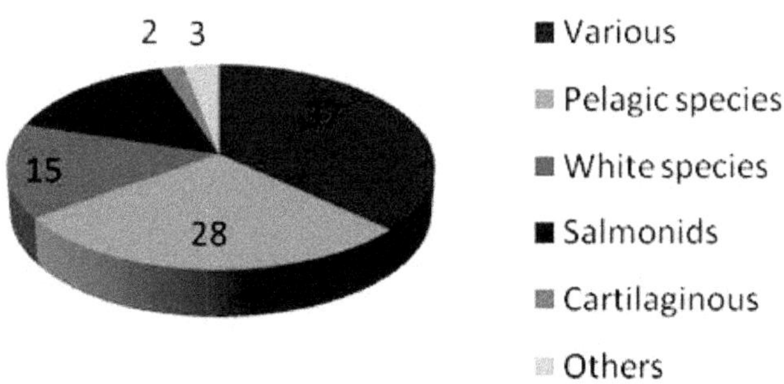

Fig. 8 Original nature of identified by-products (%), Gestion Durable project (2011).

Both methods have limitations which it is important to take into account:

- The direct method chosen by the Gestion Durable project collected real and accurate data from industry professionals as well as qualitative data and appendices, which was not possible when using the conversion coefficients concerning quantitative data exclusively. The direct method seems to be more realistic and complete regarding those aspects.

- However, it is to be noted that the enquiries are sometimes incomplete and depend on the subjectivity of the professional enquired. Also, finality is difficult to reach using such a method. There is a general lack of data, which is to be highlighted before presenting the overall results. For example, Pays de la Loire and Poitou-Charentes present weak volumes and as mentioned before, the response rate was not as good as in the other areas studied. So the results are not representative enough to come to any conclusion.

The indirect method can provide a global inventory based on information collected uniformly which allows drawing general conclusions and trends more easily, although some data do not reflect the reality.

Given these elements, it is possible to compare these two methods as each has strengths and weaknesses, which has been synthesized in Fig. 9.

Methods	Quantitative data				Qualitative data
	Reality	Accuracy	Completeness	Objectivity	
OFIMER's study	-	-	+	+	-
"Gestion Durable's" study	+	+	-	-	+

Fig. 9 Comparison of data collection methods between OFIMER and Gestion Durable.

Fish by-products Upgrading

Upgradation technologies. There are two main types of upgradation:

(a) *For mass exploitation*: by-products are used in animal feed, energy and fertilization.
(b) *For Small volume exploitation* for higher value addition for sectors such as nutrition, health food, nutraceutical, cosmetic or pharmacy.

They have different characteristics that can be synthesized using the following figure based on Maslow's hierarchy of needs pyramid. The main differences between the above two methods are the capacity of absorption by the market of the derivatives and the profit that can be derived for those who made value addition. At the top of the pyramid we find, is expected, the sectors corresponding to the small volume exploitation. Those activities do not need a lot of volumes but are more constraining in terms of quality. They require rigorous sorting, low temperatures and intensive traceability implementation.

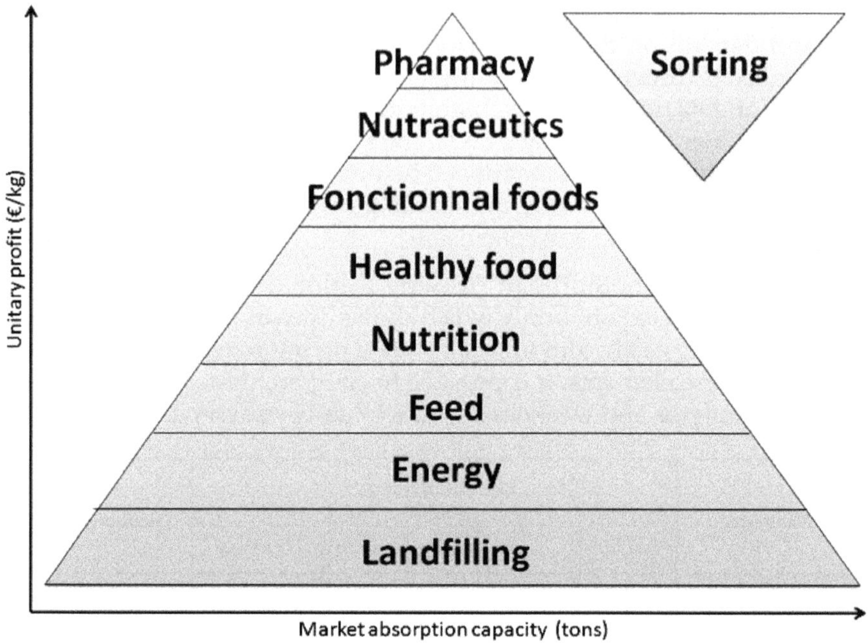

Fig. 10 Scale of marine by-products upgrading modes.

Sorting is a key factor of success to upgrade by-products for higher value products because the sectors which treat them work on specific molecules contained in the specific type of by-products (collagen in skins for instance), sometimes in specific species (e.g., Siki liver oil for aeronautics industry, etc.). The storage mode is also very important because the specifications of these activities are strict. Cold storage is a must and the collection of by-products must be done quickly. Further, the regulation for the dietetic and nutraceutical industry is especially restrictive and authorizes only a small number of products to receive food complement grades. The origin of the by-products must be formally informed to ensure a perfect traceability. For the bottom of the pyramid there are fewer constraints. The sorting is not necessary and specifications are more flexible, that is why nowadays the most important part of by-products volumes are treated this way.

Current scenario. Today by-products are generally upgraded into meals and oils, derived mainly from farm fish whose production is steadily growing in the last several years. World production of fish meal was about 5 million tons in 2008 and production of fish oil was about 1 million ton. 22 million ton of whole fish and by-products was necessary to ensure this production.

The main producers of fishmeal and fish oil are Peru and Chile, with respectively 1.4 and 0.8 million tons produced in 2007. These flours are of first quality because they are derived from whole fish (industrial fisheries). Increasingly, marine feed ingredients coming from fisheries by-products have now reached over 25% of global production (IFFO 2009).

IFFO estimated that in 2009, 63% of the global fishmeal production came from aquaculture and that was split almost equally between salmonids, marine fish, crustacean and others. 81% of global fish oil production came from aquaculture of which 68% went to salmonids.

Europe produces about 335,000 tons of fishmeal yearly and consumes about three times more.In European countries, fish meal is mainly derived from by-products because of the legislation on industrial fisheries. In France, the situation is roughly the same. The production of fish meal and fish oil is predominant with 53% of derived products coming from by-products. Animal feed constitutes 75% of the by-products and uses as petfood represent 22%. Recently, hydrolysates have become important due to diversification of upgrading activities especially in Boulogne-sur-Mer (northern France) and represent nowadays 21% (Andrieux 2004). Only 4% of derivatives are intended for high-value markets.

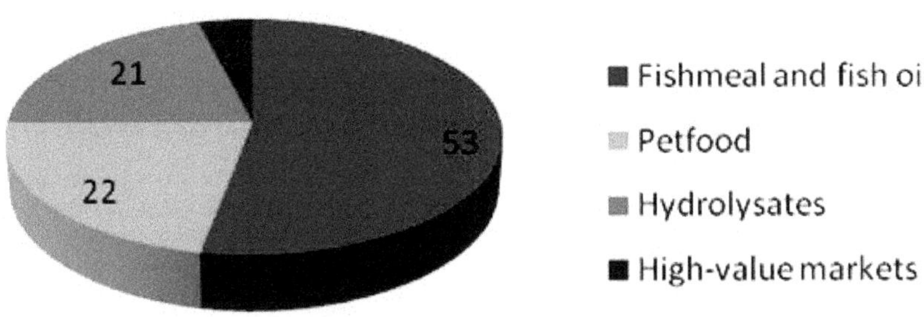

Fig. 11 Derivatives from French fish by-products (%), OFIMER (2004).

COMPARISON OF TWO FRENCH SITUATIONS

Main Processors of Fish by-products in France

There are two main processors of fish by-products in France, which work on two different territories—Copalis (initially named CTPP standing for Cooperative Processing of Fishery Products) who is the collector and processor of by-products generated in Boulogne-sur-Mer. The other one, Bioceval, collects and upgrades by-products from the entire Atlantic coast.

The modes of operation of these two companies are vastly different; their study will allow identifying strengths and weaknesses of both highlighting the trends that should be followed.[3]

The two companies present similar characteristics even if the smaller size of Copalis seems to be more important, as shown in Fig. 13 below, which compares some companies' details.

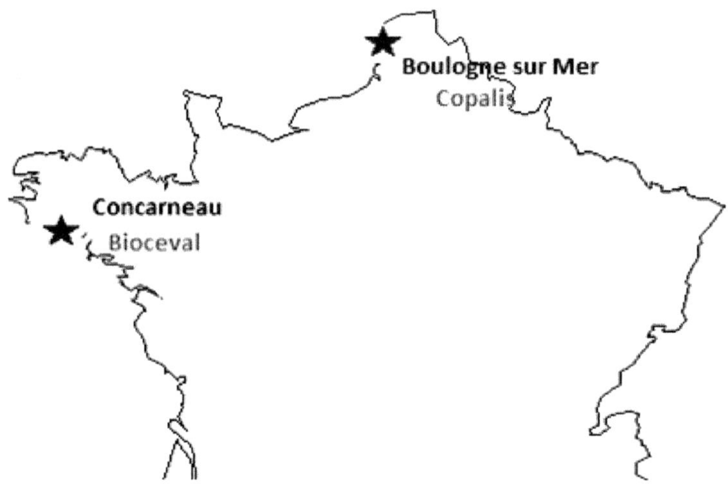

Fig. 12 Localization of the two main French fish by-products processors.

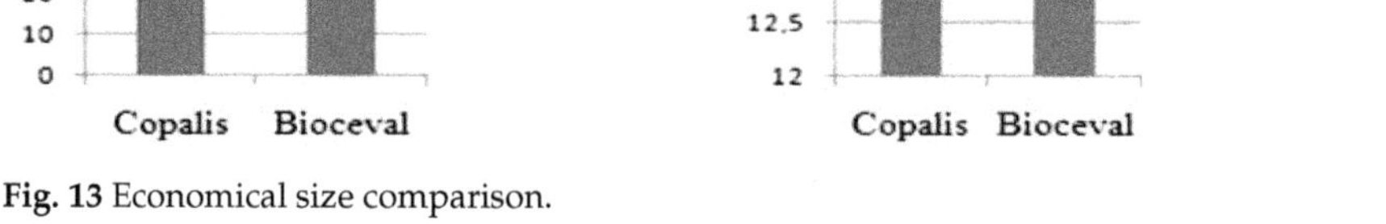

Fig. 13 Economical size comparison.

Copalis

Copalis was founded in 1960 by members of the inter-fishery of Boulogne-sur-Mer, northern France. The primary mission of Copalis is to enhance and bring value addition to by-products generated in the fishing industries

[3]The data used in this chapter are based on the personal data presented orally by Copalis and Bioceval.

of Boulogne-sur-Mer. The original producer of fishmeal and fish protein hydrolysates, Copalis diversified its business in the late 90s with the development of new technologies (fractionation and isolation of peptides) and has now firmly established itself in the development of marine bioactives.

Copalis has became a global supplier of marine ingredients to the nutraceutical, functional food, animal nutrition and cosmetics industry. Indeed, 75% of its turnover is from exports. Located in the first European center of processing, marketing and distribution of fish, Copalis is able to collect 65,000 tons of fish by-products every year in a reduced scale.

The cooperative operating method of Copalis ensures that profits are directly fed back into the company for better pay suppliers and therefore always ensures the highest quality of by-products. Such a cooperative status also allows the company to control the traceability and origin of the raw material. This is what has contributed to its success. Further, Copalis is independent unlike its French competitor Bioceval which is part of the SARIA group.In addition, Copalis features an integrated process enabling it to produce high-value added ingredients whose production usually consists of extracting one or more molecules. Its strength is that it can reintroduce the residues of these extractions which represent a large part of the volumes for the production of fishmeal or hydrolysates.

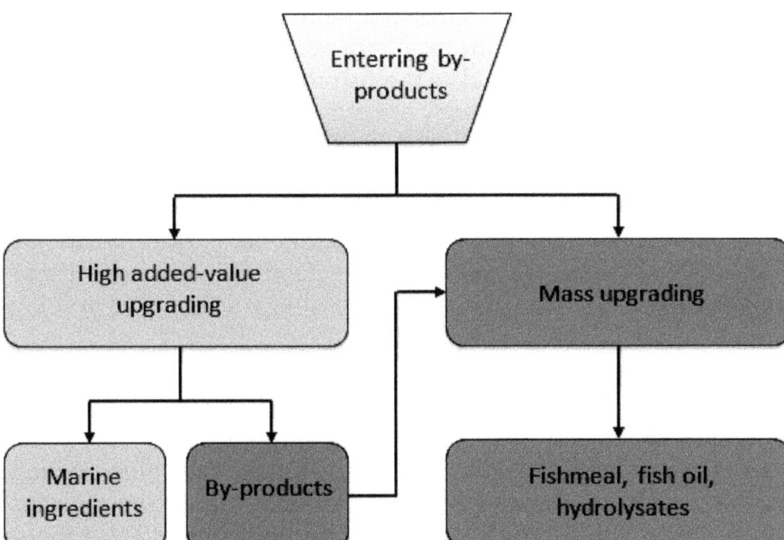

Fig. 14 Copalis' integrated process. With permission of Copalis®.

Bioceval

Bioceval, founded in 1995, is part of the SARIA Group which specializes in the treatment of numerous kinds of waste. It is the only unit in the group which works in the treatment of fish by-products to make fishmeal and fish oil destined for aquaculture. The plant is located in Concarneau in the south of Britanny. This location can be explained by the proximity of the deposits available in Bretagne. Indeed, Concarneau is located between two main places for fishing activities: Lorient and the Cornouaille region. But the activities of, Bioceval is not limited to just these two places but with the entire Atlantic coast, which represents a vast territory and involves many economic, logistical and qualitative considerations.

According to figures provided by SARIA, Bioceval collects 200 tons of marine by-products per day, representing 60,000 tons per year.

Focus on Boulogne-sur-Mer

An ideal location

At the meeting point of the fishing zones and consumer markets, the fishing port of Boulogne-sur-Mer occupies a key position in the international fisheries and seafood market of as the first French fishing port with a diversified fleet of nearly 150 boats. Boulogne-sur-Mer is central Europe's leading seafood producer with over 140 companies and over 70 different species of fish landed and sold at the fish markets here every day. 35,500 tons of fresh fish are landed directly at auctions (80 companies are approved as buyers at the fish auction) but the strength of Boulogne-sur-Mer lies in the quantity of marine products shipped. Each year, 380,000 tons of sea products pass through this center of fisheries.

The strength of Boulogne-sur-Mer is this network of businesses on the same site where all activities of the seafood industry are present and interconnected.

Fishing industry

Fishmongers are crucial players in the seafood industry, as they provide a link between production (fishing) and distribution. They prepare products in relation to market demand and expectations, and ensure their quality and traceability. The fishing industry is a fundamental pillar of the socio-economic structure of the coastal regions in France. Boulogne-sur-Mer combines 70 companies, 1,500 jobs directly related to fish industry and 400 million euros of sales on an area of less than 9 km^2.

Focus on French West Coast

Many processors scattered in space

In 2011, in the four regions studied by the "Gestion Durable" project there were 28 fish auctions, 75% of sales at French auctions and 393 fish processors. Bretagne represents the area where the fish sector is the most important, with 15 fish auctions and 50% of the quantities sold at auctions (FranceAgriMer, 2012). The scattering of those industries involves large scale collection. Indeed, taking into account the geography of the territory, the area covered by the project "Gestion Durable" is around 90,000km².

The comparison of these elements with Copalis' strategy allows to draw some conclusions. Bioceval has to collect 40 times more suppliers, a maximum distance of collecting 20 times larger and 10 times less final applications sectors, as shown in Fig. 15.

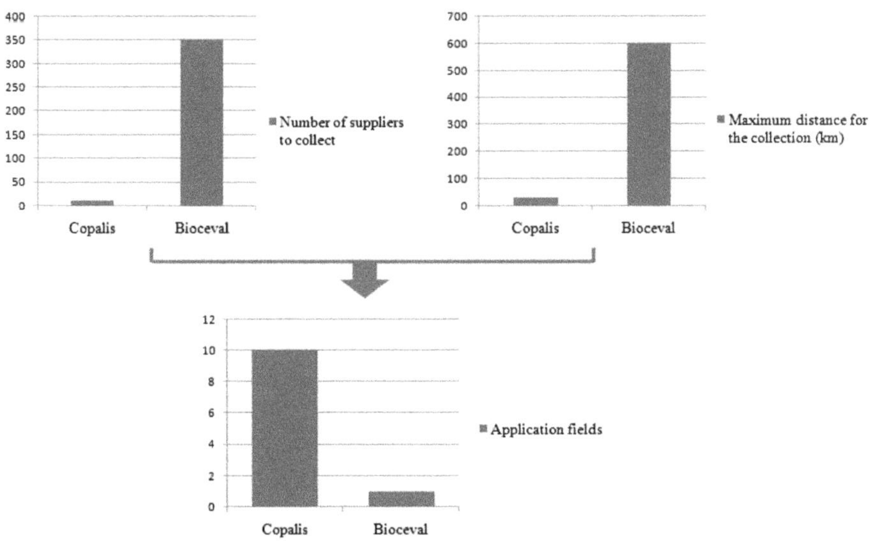

Fig. 15 Comparison between Copalis and Bioceval strategies. With permission of Copalis® and Bioceval®.

Volumes, quality, management

The results of project Gestion Durable showed that this multitude of actors scattered in space engendered complications if a federation around an upgrading project involving more complex specifications was needed. Federation of actors around a quality charter takes time and requires a relationship of trust. Too many professionals involved makes it complicated.

Indeed, management of by-products is different from one place to another. In some cases, the by-products are centralized at the auction and thereafter it is the authorities who manage the storage and choose the provider for the collection. The fixed price is then the same for all professionals, qualitative constraints as well. In other cases each professional manages its own by-products implying differences due to negotiations between industry and collection providers. These cases are distributed unevenly over the territory. Harmonization of practices is therefore difficult to implement.

Synthesis of Key Elements of Success

The detailed inventory which has been highlighted in the beginning of this paper highlights not only some blocking factors but also factors of success. These factors are related to three main issues related which are interrelated--territorial, quantitative and qualitative issues.

Territorial and quantitative parameters

Proximity of deposits and processors engender logistics issues that have an effect on several points such as prices, frequency of collection, freshness of the by-products, etc. The volumes of the deposits are another factor that has to be considered. Upgrading activities, whether mass or high value level, have important raw material quantities requirements.

Landings are not sufficient to evaluate the fish processing. In the case of Boulogne-sur-Mer, the 380,000 tons of products that pass through companies make the difference.

Qualitative parameters

Convincing the industry to respect a quality charter is a lengthy process which can facilitated when the deposits are located near the treatment sites. Exchanges between the industry actors are then more effective. Copalis successfully diversified its business by producing products with high value additions—thanks to the trust relationship that has been established between the different parts. The number of companies to federate, the proximity of the deposits and the location of the company for forty years facilitates the establishment of such upgradation modes.

Moreover, recent studies conducted in the project Gestion Durable showed that in addition to the use of chemicals and energy during the manufacturing process, transport and deterioration of by-products, which is linked to the storage mode and to the frequency of collection, have a negative impact on the environment. A greater freshness of by-products entering the process reduces ammonia emissions and the associated treatment what could allow a reduction up to 25% of some impacts on the environment, such as marine ecotoxicity, eutrophication or acidification. Those results have been carried out through the study of the life-cycle assessment of fishmeal, fish oil and hydrolysates (Cesbron 2011).

Economical parameters

Quartering was a free public service until 2005 when it was confined to dead animals' elimination. Producers of by-products had then to systematically resort to an external service to manage the collection. Before this, it did not represent any cost for professionals. Nowadays, there are private companies who are in charge and it is they who define the costs

The sorting and cold storage needed for any high-value upgradation represent significant costs for the industry, which must be offset by compensation. Logistical costs related to the collection of by-products when producers are scattered in space do not allow any remuneration. In the case of Boulogne-sur-Mer, the proximity of industries allows a remuneration which motivates producers of by-products to meet the specifications defined by the collector. In the case of the area cover by Bioceval most of the time it is the producers of by-products that must pay. The cost is fixed by the collector and depends on several factors such as:

- volume of by-products to collect
- nature of the by-products
- distance from the treatment site.

Indeed, a consequent volume of by-products available in the same site allows the collector to amortize the costs associated with transport and therefore reduce the financial repercussions on industrialists. Moreover, the nature of the by-products can reduce the cost of collection. For example, fatty fish such as salmon are used to make fish oils which are more profitable for the collector. Negotiations to reduce the cost of collection are then easier. Some companies have means to negotiate but in most cases it remains an expense or a free.

CONCLUSIONS

1. Today, the seafood sector is facing many problems on the one hand regarding the upstream activities (fishing, landings), and on the other hand concerning its downstream part (consumer expectations). Processing activities are in the middle of these changes and must continually adapt to the constraints that this entails.

2. French households consume more fish and buy it more and more transformed but at the same time landings volumes are falling. Thus, the use of import and aquaculture has increased. Imported products are mainly already processed.
3. The volume of by-products on the territory is associated with these parameters which tend to mean that they will decrease in the coming years if the situation of the fisheries sector does not improve.
4. Paradoxically, aquaculture that is becoming increasingly important, is a heavy consumer of fish meal and fish oil made from by-products and whole fish. Production from whole fish is fluctuating and subject to the same constraint that the fisheries sector especially in the producing countries where it is regulated. This suggests that the use of by-products to produce fish meal and fish oil (25% of the raw material used nowadays) will continue to rise.
5. Quantitative estimates of by-products are fluctuating as they depend on many factors that are not fixed from one year to the next: the conversion rate, imports and landings. The different methods used to estimate the volumes of by-products available in the territories highlight roughly the same trends.
6. By comparing two collectors of marine by-products operating in different territories, it is possible to highlight the success factors for the development of derivatives products for several applications including those with high added-value.
7. First, upgrading processors must work with a limited number of suppliers with whom a long-term agreement is made possible thanks to many years of presence and negotiations to ensure the highest quality of by-products. Secondly, significant volumes have to be available in the immediate vicinity of the treatment site to keep a high quality and to limit transport costs.
8. This overview of fish by-products upgrading is aimed at determining what could be optimized and to find complementary alternatives in a context of sustainable development. These alternative solutions may be found at a larger scale by including other kinds of organic by-products. For territories where the volumes of fish by-products are not sufficient to be treated alone and where at present they have to travel hundreds kilometers to reach their treatment site, it could be considered to upgrade them in a generic process at the local or the regional level with by-products from many other sectors (vegetables, meat, poultry, cereals).

REFERENCES

ADEME. 2010. Biodéchets du commerce et de la distribution.

Andrieux, G. 2004. La filière française des co-produits de la pêche et de l'aquaculture, état des lieux et analyse.

Cayeux, S. 2007. Les français à la recherche du prix et de la praticité. Panel de consommateurs TNS Worldpanel, présentation des données arrêtées au 31 aout 2007.

Cesbron, E. and A. Cikankowitz. 2011. Évaluation environnementale de filières de valorisation des sous-produits de la mer, Projet PSDR, Grand Ouest, Série Les Focus PSDR3.

COGEPECHE. 2009. Comportement, besoins et attentes des consommateurs.

Dumay, J. 2006. Extraction de lipides en voie aqueuse par bioréacteur enzymatique combiné à l'ultrafiltration: application à la valorisation de co-produits de poisson (*Sardina pilchardus*). Ph.D. Thesis.

European Commission. 2007. A policy to reduce unwanted by-catches and eliminate discards in European fisheries. COM (2007) 136. Commission of the European Communities, Brussels.

European Commission. 2011. Proposals for reform of the Common Fisheries Policy. Commission of the European Communities, Brussels.

FAO. 2004. Discards in the Word Marine Fisheries: an update. Food and Aquaculture organization, Rome, Italy.

FAO. 2006. The State of Word Fisheries and Aquaculture. Food and Aquaculture Organization, Rome, Italy.

FAO. 2010. The state of Word Fisheries and Aquaculture. Food and Aquaculture Organization, Rome, Italy.

FAO. 1994. A global assessment of fisheries bycatch and discards. Food and Aquaculture organization, Rome, Italy.

FAO. 1995. Code de conduite pour une pêche responsable. Food and Aquaculture organization, Rome, Italy.

FranceAgriMer. 2009. Consommation des produits de la pêche et de l'aquaculture.

FranceAgriMer. 2010. Consommation des produits de la pêche et de l'aquaculture.

FranceAgriMer. 2012. Chiffres clés de la filière pêche et aquaculture.

IFFO. 2011. Fishmeal and Fish Oil—The Facts, Figures, Trends. International Fishmeal & Fish Oil Organisation.

Le Floc'h, P., P. Bourseau, F. Daurès, F. Guérard, L. Le Grel, M. Meunier and M. Tuncel. 2011. Valorisation des co-produits de la mer et territoire: enjeux territoriaux. Revue d'Économie Régionale et Urbaine. 1: 213–225.

OCDE. 2012. Allègement des taxes sur le carburant. Comité des pêcheries.

Pérez-Gálvez, R. 2010. Management of fish discards and by-products on board. LAP Publishing, Saarbrücken, Germany.

2

Improved Utilization of Fish Waste, Discards, and By-products and Low-value Fish towards Food and Health Products

Carlos Cardoso and *Maria Leonor Nunes*

ABSTRACT

Fish food is a staple diet product in many countries. Indeed, fish products are important nutritive, and healthy food. Precious nutritive components are found in fish, mainly, vitamins, minerals, proteins with a balanced amino acid profile, and omega 3 polyunsaturated fatty acids (ω3-PUFA). Some of these nutrients are considered functional ingredients, given their positive effect in the regular physiological functioning of the human organism and, also in disease prevention. Accordingly, the growing concern of consumers with food safety and health, has led to a higher demand for fish products. As a result of overexploitation and environmental changes, the world catch have not been able to keep pace, thus leading to an increase in aquaculture production. However, other solutions for the overexploitation problem

National Institute of Biological Resources, INRB, I.P./L-IPIMAR, Avenida Brasília, 1449-006 Lisboa, Portugal.

may be found. There are underexploited resources, low-value fish and fish waste, discards, and by-products. The utilization of these resources is a great challenge for the sector, requiring innovative technological strategies. Many solutions can be obtained from restructuring fish products and extraction, and isolation of relevant components, such as lipids, proteins, enzymes, peptides, chitin, and chitosan. In this way significant amounts of undervalued fish materials are upgraded and find application as foods or nutraceuticals. However, the novelty of these products in the context of a rapidly changing landscape, particularly the nutraceuticals, raises new questions related to the adaptation of the legal regulation that frames production, quality, and marketing of these products.

INTRODUCTION

Fish is the staple food in many countries. Indeed, fish products are a very important source of nutritive, pleasant, and healthy food. Precious nutritive components are found in fish, mainly, vitamins, minerals, proteins with a balanced amino acid profile and it is readily digestible (Borderías et al. 2005), full of omega 3 polyunsaturated fatty acids (ω3-PUFA), particularly eicosapentaenoic acid or EPA (20:5ω3) and docosahexaenoic acid or DHA (22:6ω3). Some of these nutrients are considered functional ingredients, given their positive effect in the regular physiological functioning of the human organism and, even in disease prevention. In fact, several epidemiological studies (Simopoulos 2002) have found a reduced risk of coronary heart disease, hypertension, stroke and consequent further effects due to the beneficial effects of ω3-PUFA on lipids, namely cholesterol, blood pressure and eicosanoid levels as well as on the coagulation rate. Hence, for the primary prevention, the American Heart Association (AHA) recommends a daily intake (RDI) of 500 mg EPA+DHA (Kris-Etherton et al. 2002).

The growing concern of consumers with food safety and health has led to a higher demand for fish products. However, as a result of overexploitation and environmental changes, world captures have not been able to keep pace with this demand, thus leading to an increase in aquaculture production. World marine captures may be nearing the catch ceiling for commercial resources. There are widespread concerns about collapsing stocks, over-capacity in fishing fleets, unsustainable fishing practices, fish discards (in some fisheries the discards can be up to 40–50% of the catch) wastage, of fish processing by-products and all the consequent environmental degradation. Indeed, the fish processing industry generates significant amounts of waste and low-value by-products that if not upgraded may lead to high disposal costs. In some areas the only practical option is to send the waste to landfills. However, aquaculture production may be only one possible solution to these problems. In fact, there are large quantities of fish of low commercial value which are treated as waste or used to produce fish meal. Hence, there are underexploited resources, low-value fish and fish waste and by-products from the fish processing industry which are deemed unmarketable for various reasons, such as unpleasant taste, too soft texture, are bones, have a high fat content or small two small in for human consumptions. These are however, valuable resources from a nutritional point of view, containing proteins, vitamins, minerals, and ω3-PUFA as good as those contained in the commercial products, provided that adequate handling and storage conditions are ensured, taking into account hygienic conditions.

Upgrading of underutilized fish species and by-products of the fish processing industry is a fundamental difficulty for the sector, since it requires adequate and innovative technological strategies to be overcome. Important technological answers are offered by the preparation of restructured fish products and the extraction and isolation of relevant components, such as lipids, proteins, enzymes, peptides, chitin, and chitosan. In this way undervalued fish materials are upgraded and find application as functional foods or nutraceuticals.

Restructured fish products are foods made from minced and/or chopped fish muscle, which are used, with or without ingredients, to make products with a new appearance and texture (Borderías et al. 2005). The rationale behind restructuring fish muscle is that the supply of high quality fishery products is limited and many are becoming exhausted because of excessive fishing. One of the chief advantages of restructured products is that the composition of the end product can be modified by reformulation of the original product once it has been chopped or ground. In this sense, the process might be said to be one of eliminating some constituents or adding other new ingredients or additives. This approach may enable to remove, offset or, at least, mitigate the deleterious properties of some low-value fish and fish waste and by-products.

On the other hand, recent studies have identified a number of bioactive compounds from fish muscle proteins, collagen and gelatin, fish oil, fish bone, internal organs and shellfish and crustacean shells (Je et al. 2005, Jeon and Kim 2002, Kim et al. 2001). Generally, a more profits may be obtained by producing human consumables. These bioactive compounds can be extracted and isolated with technologies varying from simple to complex and such processes may include preparation and isolation of bioactive peptides, oligosaccharides, fatty acids, enzymes, chitin, water-soluble minerals, and biopolymers for biotechnological and pharmaceutical applications. Some of these bioactive compounds have been identified to possess nutraceutical potentials that are beneficial to human health (DeFelice 1995). Accordingly, development of new technologies in search of novel bioactive compounds from low-value fish and marine processing by-products may bring more value out of what is today considered a waste and presents unique challenges and opportunities for the seafood industry (Kim and Mendis 2006). This chapter will cover a wide group of products and technologies, but it will not deal with fish hydrolysates, which has been discussed in a separate chapter.

MINCED FISH PRODUCTS

Restructured fish products encompass a large range of foods right, from fish fingers, which are specially targeted for children (breaded, frozen, and to be used as fried products) to a new generation of products called analogues or substitutes, most of them mimicking traditional seafood products or, even, other high value products. These have not only become popular in Japan and other countries of the Asia-Pacific area, but have also gained a market share in North America and, more recently, in Europe. The vast majority of these products are prepared from surimi, which is thoroughly washed and refined fish mince (Borderías et al. 2005). However, there are products prepared directly from minced fish—washed or not washed—and others from fish protein isolates or recovered proteins (Fig. 1).

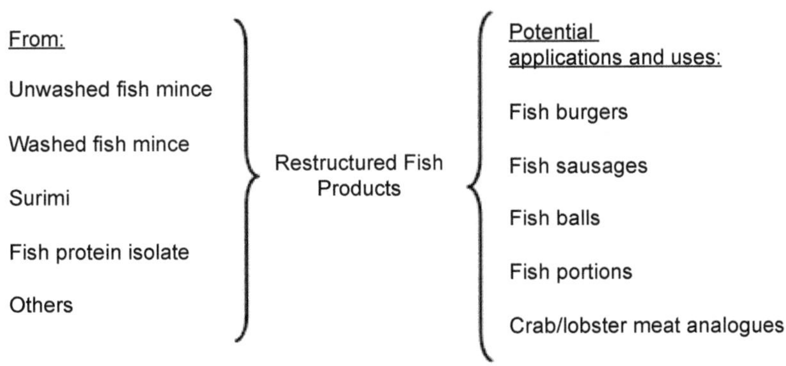

Fig. 1 Categories and examples of restructured fish products.

Another major division is between gelled fish products and other food groups. The former are presented in Chapter 4. The latter comprise a wide range of foods, from fish burgers to sausages. These are attained by the mixture of fish mince (or recovered fish protein) with high amounts of fat, requiring the incorporation of emulsifier proteins and belong to the group of emulsion fish products.

In particular, fish mince can be defined as deboned and unwashed fish flesh. Minced fish contains all the proteins and biomolecules of the fish muscle, i.e., besides structural proteins, all the enzymes. It is prepared from the industry by-products and low-value fish species and can be an inexpensive material for manufacturing various seafood products (Kim and Park 2007). Moreover, as a result of a low level of food processing (no washing steps), it offers nutritional advantages (containing water-soluble vitamins, minerals, and lipids) as well as functional advantages (exhibiting meat-like texture) compared to the other intermediate materials (Lanier 1994). Additionally, higher yields in fish mince (unwashed) are especially important as waste associated with the manufacturing process is reduced. However, the conversion of fish mince to seafood final products with functional properties (from an organoleptic or health point of view) requires the addition of enzyme inhibitors or rapid cooking in order to minimize quality deterioration by enzymatic activity. Usually, the external appearance of mince from filleting by-products is poor due to the presence of pigments mixed with the flesh during mechanical separation. This problem can be overcome by the removal of the backbone prior to separation, whitening or masking. For whitening, titanium dioxide and hydrogen peroxide have been extensively used, but the results are not totally satisfactory (Taskaya et al. 2010). Masking is, in principle, a good alternative if suitable final products allowing for the use of masking agents are identified.

Diverse final products have been prepared from fish mince. From simple fish fingers or fish burgers and portions to other products involving the processing of fish mince in a way that mimics muscle fibres, resulting in fillet analogues (Moreno et al. 2008, Moreno 2009). For all these products, the addition of salt is essential as it is required for the extraction and solubilization of the muscle fibre (myofibrillar) proteins (Kuraishi et al. 1997). An important example of the utilization of fish mince in the preparation of restructured fish food is the striped mullet (*Mugil cephalus*) (Ramírez et al. 2007a). This species is found in abundance in Mexican waters. However, it is mainly captured to commercialize its roe as the flesh has low commercial value. Incorporation of additives (amidated low methoxyl pectin) was required for achieving products with adequate mechanical properties (Ramírez et al. 2007a). Furthermore, in this case, it was also possible to lower the salt content (to 1.0%, w/w) through microbial transglutaminase (MTGase) or whey protein concentrate addition (Ramírez et al. 2007b).

In this context, with the purpose of upgrading hake (*Merluccius capensis*) trimmings from the cutting operation in the frozen hake industry, a fish sausage was developed (Cardoso et al. 2008a). The textural and sensory properties of this sausage were very similar to commercial pork sausage. Again, this was possible through the incorporation of functional ingredients, namely, pea and chicory dietary fibres (DF). Both fibres ensured a significant portion of the daily DF requirements, thus increasing the nutritive value of the sausage. In particular, pea DF offset the softening effect of replacing pork meat and fat by hake mince and chicory DF was quite effective in simulating the creamy feel typical of fatty products (Cardoso et al. 2008a). Accordingly, a hypocaloric pasteurized hake sausage was produced whose sole fat source was hake, a lean fish species with a high proportion of ω3-PUFA in its fatty acid profile. Besides, this product displayed a chill storage stability of about 45 d at 2 ± 1°C (Cardoso et al. 2008b). However, the fish species was influential on sausage quality and the utilization of other raw materials did not enable results as favourable as with hake (Cardoso et al. 2009a). With a similar goal, a minced hake product with shellfish flavour, enriched in ω3-PUFA, and intended to be served as an appetizer was also developed (Cardoso et al. 2010). This novel food was favourably assessed by a sensory panel and exhibited no quality and safety threatening changes over a period exceeding 3 months (Cardoso et al. 2010).

Besides hake, by-products of other fish species processing have been used in the preparation of novel products. For instance, the filleting wastes (trimmings and backbone) from silver catfish (*Rhamdia quelen*) were minced and used in the formulation of fish burgers (Bochi et al. 2008). It was found that fat and moisture retention after cooking as well as texture and juiciness were not affected by the level of waste incorporated into the product. Despite burgers containing 80% waste and 20% fillet having lower overall acceptance than controls with no waste, results indicated that production of fish burgers may be a valuable alternative for wastes from silver catfish filleting. Filleting wastes could substitute up to 50% of fish fillets with no changes in sensory acceptance and an improvement in nutritional value (higher ω3/ω6 ratio) and cooking characteristics (Bochi et al. 2008).

Therefore, minced fish products may offer an outlet for low-value fish and fish by-products, being possible the formulation of high added-value functional foods through a proper storage and handling of the raw material and the utilization of suitable additives (Blanco et al. 2007).

GELLED PRODUCTS

This class of fish products also comprises a great variety of foods and different raw materials, but what all these products have in common is the gelling phenomenon. Gelling is characterized by the formation of a three-dimensional protein network between the myofibrillar proteins, thereby leading to specific viscoelastic properties (Meyers and Chawla 1998, Lanier et al. 2005) (Fig. 2).

These proteins have highly reactive surfaces once the protein is unfolded. During heating of fish mince (washed or not), proteins unfold, exposing their reactive surfaces to the neighbouring protein molecules, which then interact to form intermolecular bonds. For this purpose, salt is fundamental (Kuraishi et al. 1997). Usually a decrease in salt level to below 2.0% (w/w) has a negative effect on the mechanical properties. In fact, some

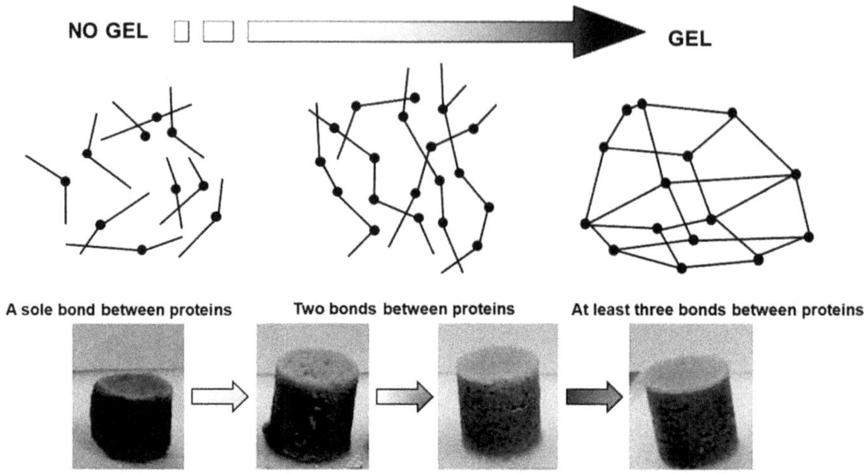

Fig. 2 The chemical bonding between proteins and its relation to the gelling process.

authors (Gómez-Guillén et al. 1997) proposed a minimum level of 1.5% (w/w) salt. Nevertheless, low-salt gelled fish products with MTGase as the binding agent were obtained using mechanically deboned fish meat from filleting by-products of silver carp (*Hypophthalmichthys molitrix*) (Téllez-Luis et al. 2002). These authors demonstrated the feasibility of attaining low-salt (1.0%, w/w) gelled fish products with improved mechanical properties from silver carp mince.

Within gelled fish products, an important group is prepared from washed fish mince and within this group of raw materials, from surimi. The surimi processing method was originally developed in Japan (on the basis of traditional ancient processes). In fact, surimi is a word originally coined in Japan, meaning minced muscle. To overcome theproblem of low frozen storage shelf life (Yoon et al. 1991). The minced flesh is repeatedly washed with chilled water (<10°C) or with saline solutions containing sodium bicarbonate or sodium chloride, which removes the unnecessary components that promote protein denaturation, particularly during the frozen storage (Park and Lin 2005) (see Fig. 3).

Before 1960, surimi was manufactured and used up within a few days as a refrigerated raw material, because freezing induced protein denaturation. However, with the discovery of cryoprotectants, the surimi industry was able to tap into previously unexploitable resources (Park and Lin 2005). Since then the surimi industry has greatly expanded, not only in Japan, but also in the USA and more modestly in Europe (Kim and Park 2007). In general, it is considered that surimi has excellent gelling and water-binding properties (Lanier 1986). The high concentration of myofibrillar protein in surimi enables the product to gel upon heating to produce a chewy and elastic texture (Bertak and Karahadian 1995).

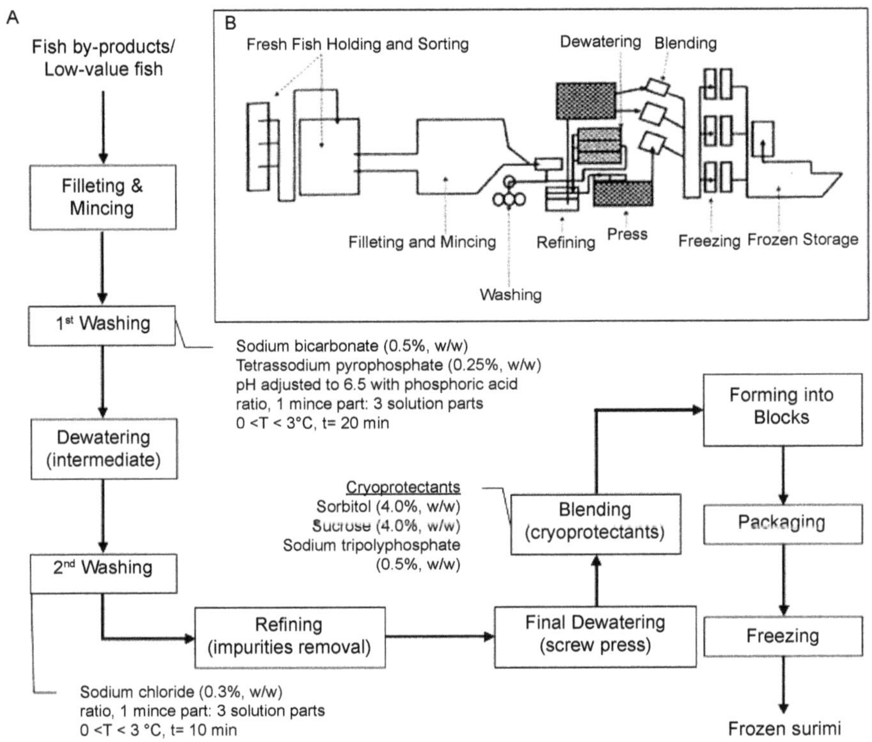

Fig. 3 Main steps of a typical process (A) and flow chart (B (adapted from: Draves 2003) of surimi manufacturing.

Therefore, through a simple technology it is possible to produce seafood products of various types and flavours, such as lobster or crab meat analogues (Babbit and Reppond 1987). Worldwide, crabsticks have been one of the most prevalent gelled seafood products in the marketplace (Hollingworth et al. 1991).

As mentioned earlier, with regard to minced fish products where gelling is of secondary importance (Chapter 3. Minced fish products), surimi has also been produced from underutilized fish resources. Regarding small pelagic fish underutilized by the industry, some studies (Bentis et al. 2005) have shown the feasibility of preparing the final products with adequate textural quality (hardness and elasticity) and high nutritive value (ω3-PUFA). This has been part of a wider effort to exploit some underutilized fish species, given the limited access to the Alaska Pollock sources (Chen et al. 1997), the main raw material for the production of surimi (Guenneugues and Morrissey 2005). However, despite serious attempts to use underutilized small pelagic species, such as mackerel, the control of the factors influencing the large-scale processing of such raw materials has remained difficult (Chen et al. 1997). Namely, the factors such as high fat content or the pigment and trimethylamine oxide amounts in dark-fleshed raw material. Consequently, different and more adequate washing procedures were required in order to prepare high quality surimi from such raw material (Shimizu et al. 1992). For instance, improvements in the colour of surimi with a higher number of washing cycles (Kim et al. 1996) or as a result of different washing times and water quantities (Chen et al. 1997) have been reported. An example of the utilization of a low-cost species for the preparation of surimi is the silver carp (Luo et al. 2008). Even more recently, some significant progress was achieved regarding the preparation and utilization of surimi from low-value small pelagic fish through the simultaneous addition of MTGase and DF (Cardoso et al. 2009b).

Surimi however, has some drawbacks including low recovery yields, loss of some nutritionally valuable components, and the production of large quantities of wastewater, which are a serious environmental problem (Kim and Park 2007). Hence, unwashed fish mince may be an alternative intermediate material suitable for the utilization of industry by-products and low-value fish species. MTGase and DF have also been applied as enhancers of the gelling ability of unwashed fish mince (Cardoso et al. 2007a,b). Generally, MTGase was more successful than DF, clearly improving gel strength with incorporation levels of just 0.5% (w/w) (Cardoso et al. 2007a). Nevertheless, DF enhances the functional value of the products and may generate texture profiles adequate for novel products such as patties from hake trimmings (Cardoso et al. 2007b). More recently, trimmings (muscles joined to the bones and skin) of the processing of farmed fish, such as sea bass (*Dicentrarchus labrax*), have been minced and successfully used in the preparation of gelled products (Cardoso et al. 2011). Their textural quality was much improved by the combination of two DFs—carrageenan (1%, w/w) and konjac glucomannan (1%, w/w). Particularly, these DFs exhibited a strong hardening effect on the gels. The tested technological solutions suggest that for commonly farmed fish species, whose market for whole fish presents signs of saturation, thereby creating a potential opportunity for ready-to-cook fillets, the concomitantly generated by-products may find an outlet as human food (Cardoso et al. 2011). Hence, it is possible that in future, crab sticks, 'angulas' (gelled surimi moulded to mimic elvers, young eels), and other gelled products be produced from low-value fish species and various fish by-products through the application of MTGase, DF, and other additives (Cardoso et al. 2012).

RECOVERY OF PROTEINS AND LIPIDS

As an alternative to the preparation of restructured fish products, the extraction and isolation of relevant components, such as lipids and proteins, can be a viable approach (Hultin et al. 2005). This can be partially justified by the contamination of trimmings with other tissue components such as blood and sarcoplasmic proteins, which may jeopardize the textural quality of the restructured products, thus requiring balancing or mitigating additives. On the other hand, much higher added-value can be created by the extraction and isolation of relevant components. Some of the problems associated with the use of trimmings and other by-products are similar to those encountered with the small pelagic dark-muscled fish species, which constitute a very significant portion of the underutilized fish (Hultin et al. 2005).

In this context, the advantages of a process for isolating a high-quality human food not only from fish by-products, but also from low-value fish, are obvious. Accordingly, a process for attaining functional protein isolates from by-products and dark-muscled fish species has been developed (Hultin and Kelleher 1999, 2000). The overall process concept is straightforward and it is known as the pH shift process (Fig. 4). The proteins of the muscle tissue are first solubilized. This is accomplished by mixing 5 to 10 volumes of water with alkali added to obtain approximately a pH of 10.5 or higher, or with acid added to about pH 3.5 or lower. The mixture is then centrifuged. This allows the light oil fraction to rise to the top of the suspension. Simultaneously, the membrane lipids are removed as sediment. Other insoluble impurities, such as bone or skin, are also gathered in the sediment. Afterwards, the soluble protein fraction is precipitated by adjusting the pH to a value near the isoelectric point of most proteins, that is, in the pH range 5.2–5.5. The non-protein soluble components from the muscle tissue remain in the supernatant fraction after centrifugation and are separated from the protein-rich fraction (Hultin et al. 2005).

The pH shift process has been used in the isolation of proteins from different underutilized fish resources. This has been done the help with of continuous centrifugation, for instance, using rainbow trout (*Oncorhynchus mykiss*) processing by-products (Chen et al. 2007, Chen and Jaczynski 2007). Meanwhile, a continuous bio-reactor system for fish processing byproducts has been proposed (Torres et al. 2007). The bio-reactors are equipped with built-in pumps for various processing additives such as antifoams, coagulants, flocculants, etc. The flocculants enable more efficient separation of the precipitated fish muscle proteins from the process water and thus facilitate process scale-up from a laboratory to pilot/industrial scale (Taskaya and Jaczynski 2009).

For the application of this new methodology, the protein recovery yield is a critical parameter. Since it also allows recovery of fish oil, determination

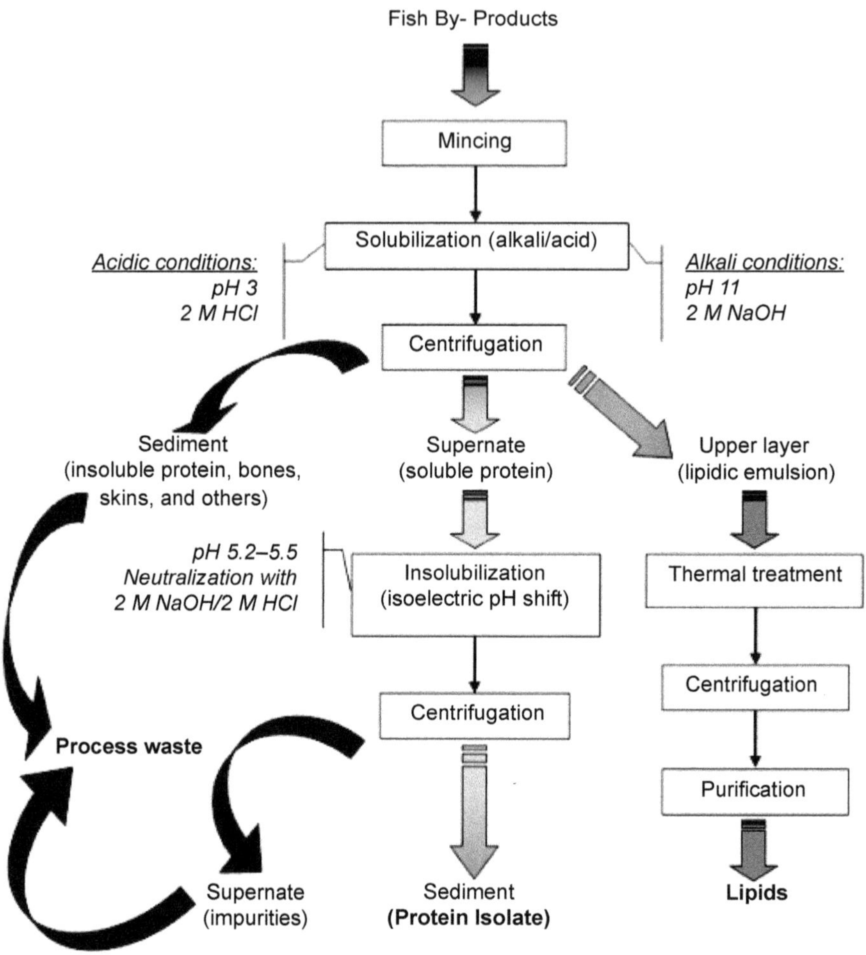

Fig. 4 Scheme of pH shift process for isolating protein and lipids from fish muscle.

of fat recovery yield may help in a more comprehensive understanding of the economical feasibility of this technology (Gehring et al. 2011). Concomitantly, fish oil that is retained with the recovered protein may contribute to the development of rancidity, thus maximal separation of the protein and lipid fractions is desirable. Besides, unlike surimi technology, the isoelectric processing allows processing of by-products such as frames and heads, thereby helping to augment the protein recovery yield. If this processing is designed as a continuous bio-reactor system with re-cycling of the process water, much less water is needed than with other strategies for the upgrading of fish by-products and low-value fish (Torres et al. 2007).

Various studies have shown that the pH shift process results in relatively higher recovery of proteins from the starting materials and alkaline solubilization usually results in higher protein recovery levels than acidic treatments. Protein recovery yields for isoelectric processing of fish by-products and low-value fish species reported in the literature range is between 42% and 90% (Chen and Jaczynski 2007, Chen et al. 2009, Kristinsson and Liang 2006, Nolsoe and Undeland 2009, Taskaya et al. 2009). The differences may be ascribed to different methods used to determine protein concentration and recovery yield, fish species, centrifugation force used during separation steps, and the relative concentration of water-soluble sarcoplasmic proteins. In fact, sarcoplasmic proteins are only partly recovered with isoelectric processing (Chen and Jaczynski 2007). Similar to protein recovery yield, fat recovery is more efficient when the pH shift is conducted at alkaline rather than acidic conditions. It has also been reported that fat recovery is augmented when citric acid and its Ca salts are used during processing (Liang and Hultin 2005).

Regarding yields, it has been found that fat recovery yield ranged from 80 to 95% for whole gutted carp processed by the pH shift process (Taskaya et al. 2009). Other authors have calculated similar fat recoveries from catfish fillets (Kristinsson et al. 2005).

Many published works have suggested that isoelectric processing of fish at alkaline pH allows the recovery of proteins with generally better functionalities as it leads to better thermally induced gelation and texture as well as better colour (whiteness) properties (Chen and Jaczynski 2007, Kristinsson and Hultin 2003, Kristinsson and Liang 2006, Nolsoe and Undeland 2009). On the other hand, degradation of myofibrillar protein was observed in acid-isolated protein (Rawdkuen et al. 2009). Furthermore, pH shift to alkaline conditions allows the recovery of muscle proteins with higher nutritional quality as assessed by a greater amount of essential amino acids compared to processing at acidic pH (Chen et al. 2007, 2009).

Compared with other protein sources, recovered fish proteins display some advantages, but also drawbacks. For instance, protein isolate recovered by alkaline solubilization from hake cut offs and "sawdust" had higher fat absorption capacity, lower emulsifying and foaming capacity than soy protein concentrate, pea protein isolate, and egg white powder (Pires et al. 2012). The gel strength of the isolated protein was intermediate (44.8 N.mm), increasing with the addition of MTGase (Pires et al. 2012). The relative advantages of protein isolates have led to their application in restructured fish products together with fish mince. Namely, it was found that Frankfurter fish sausages with 20% isolated protein (from hake by-products) substitution displayed acceptable textural and sensory properties (Pires et al. 2009). Another example is the development of a functional food product prepared from isolated rainbow trout protein (bone-in, skin- and scale-on), which was nutritionally enhanced with ω-3 PUFA and whose colour and texture properties were acceptable (Tahergorabi et al. 2012).

Isoelectric processing allows efficient recovery of the functional and nutritious proteins and lipids from sources difficult to process through conventional means. The high efficiency possible with this process may yield proteins with increased functionality and oxidative stability over conventional protein concentrates/isolates or surimi (Gehring et al. 2011). A thorough understanding of muscle proteins and lipids as well as their isoelectric behavior in food systems will enable the development of this technology at a commercial scale for efficient preparation of innovative foods (Gehring et al. 2011). Nonetheless, this requires the development of final applications for the recovered materials, a problem whose solution has become nearer with recent applied research work (Pires et al. 2009, Tahergorabi et al. 2012).

Another difficulty with isoelectric processing is the huge water consumption due to the sequential solubilization steps. Hence, life-cycle assessment studies on this process are warranted and should lead to a critical appraisal of the net worth of using this process for upgrading by-products instead of other solutions (such as preparation of products from unwashed fish mince).

With respect to lipid recovery, most of the usual processes do not involve the pH shift strategy, being composed of two main steps—oil extraction from the raw material and refining. Some other new processes to obtain ω3-PUFA concentrates for pharmaceutical or nutritional purposes include the enzymatic methods and methods that apply supercritical fluid technology (Rubio-Rodríguez et al. 2010).

The most common process to obtaining crude fish oil at industrial scale is the wet pressing method, as described by FAO (FAO 1986) (Fig. 5). This process comprises the production of fish oil and fish meal through several steps, namely, cooking of the raw material, pressing of the cooked material and finally filtration or centrifugation to recover the oil from the micella. Regarding applications to fish waste and by-products, a study has been published about the separation of oil from tuna heads by a similar method (Chantachum et al. 2000). On the basis of the yield and the quality of the oil extracted from the pre-cooked (100°C for 60 min) and non-precooked tuna heads at different heating temperatures and times, it was concluded that the optimum results were achieved when heating at 85°C for 30 min without precooking. Moreover, it has been proposed oil production from herring by-products using a method consisting of mincing the raw material (16 mm), cooking it in a heat exchanger (8 min at 95°C), and recovering the oil in a three-phase decanter (4 min at 95°C) with acceptable results (Aidos et al. 2001). These authors also studied the influence of by-product freshness (Aidos et al. 2003a) and of different process parameters (cooking temperature, pumping speed to the heat exchanger, and decanter speed) on fish oil quality (Aidos et al. 2003b). It was observed good storage stability for herring by-products and the separation speed of the decanter and the pumping speed were the main variables affecting oil quality.

Among the most innovative oil extracting processes, the enzymatic method can be mentioned as a method of releasing lipids from fish waste and by-products avoiding the use of solvents and high temperatures (Rubio-Rodríguez et al. 2010). The enzymatic hydrolysis of salmon frames

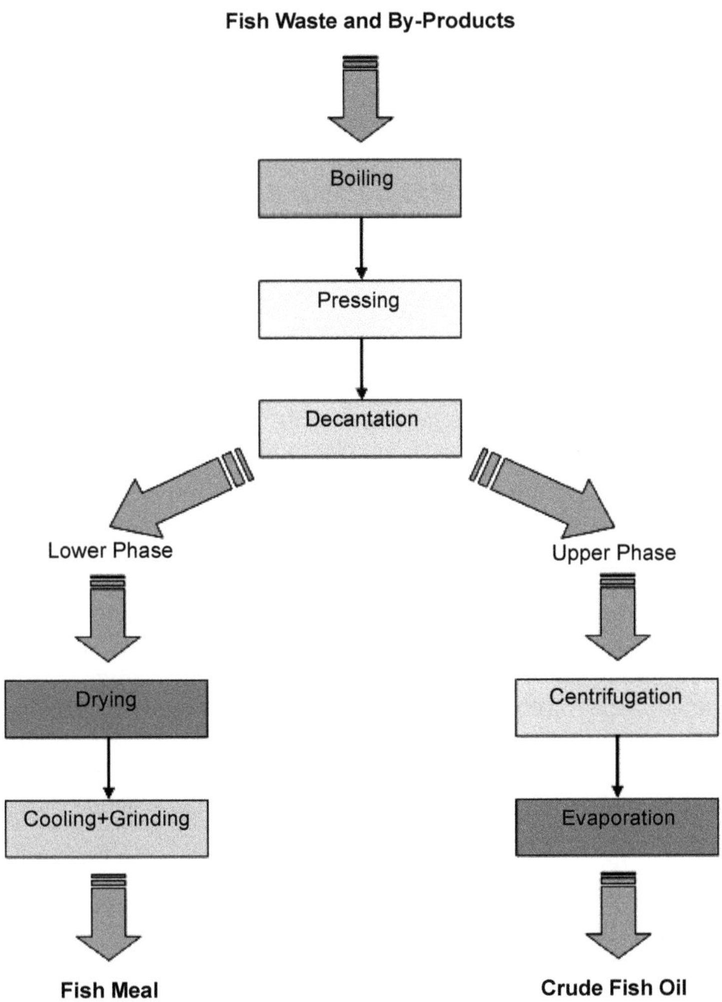

Fig. 5 Flow diagram of the usual method to obtain crude fish oil (adapted from: FAO 1986).

with proteases and the composition of the different fractions obtained after separation by centrifugation were studied (Liaset et al. 2003). This process enabled to attain ω3-PUFA enriched oil with a good recovery (about 77%). Other authors developed another enzymatic method to extract oil from ground salmon heads at middle temperature (55°C) using different commercial enzymes: a protease (Alcalase), an exopeptidase (Neutrase) and an endo-peptidase (Flavourzyme) (Linder et al. 2005). They concluded that the highest oil recovery (17.4% after 2 h) was obtained by using Alcalase. Moreover, the enzymatic hydrolysis of mixtures of cod by-products using Flavourzyme and Neutrase has been tested (Šližyte et al. 2005, Daukšas et al. 2005). They studied the influence of the composition and state of the substrate on the yield and composition of the hydrolysis products (oil fraction, hydrolysed protein, sludge, and emulsion) (Šližyte et al. 2005) and on the chemical composition of the lipid phase (Daukšas et al. 2005), being derived optimal conditions. The general conclusion reached by the several authors who studied fish oil extraction through enzyme technology is that this technology yields oil of better quality than thermal treatments (Rubio-Rodríguez et al. 2010).

ENZYMES

Fish and shellfish waste as well as low-value fish contain a wide variety of genetic material and, hence represents an enormous potential for discovering different enzymes (Raa 1990). On the other hand, nowadays, enzyme-based technology represents an important contribution (and, in some cases, essential one) for many industrial applications (Rasmussen and Morrissey 2007). This is because enzymes possess a high selectivity, and a very high activity even at very low concentrations and mild conditions of pH and temperature, which leads to fewer unwanted side-effects and by-products (Shahidi and Kamil 2001). The internal organs of fish and the shells of crustaceans constitute natural sources of enzymes with huge interest. Due to the prevailing environmental conditions, marine enzymes can effectively operate at low temperatures (below 4°C), and within neutral to alkaline pH values (Ferraro et al. 2010). Hence, the extraction and isolation of enzymes may constitute a valid strategy for the upgrading of low-value fish species and by-products of the fish processing industry.

Proteases (enzymes that catalyse the hydrolysis of peptide bonds) constitute at present the dominant group of marine enzymes which are commercially viable. Gastric, intestinal, and hepatopancreas proteinases (acting on proteins) have been studied more thoroughly (Beynon and Bond 2001): pepsin from polar cod is one of the most extracted gastric proteases, followed by collagenase, elastate, trypsin, and chymotrypsin, as well as non- proteolytic enzymes, such as transglutaminase, lipases, and chitinolytic enzymes to lower extents (Rasmussen and Morrissey 2007). The digestive organs of viscera have been investigated to see the possibilities offered by enzymes, mostly proteases (Batista and Pires 2002). A suitable method for large-scale recovery of proteolytic enzymes from fish viscera silage has been tested (Reece 1988).

Several enzymes have been isolated from fish and shellfish processing discards yielding, alkaline phosphatase, hyaluronidase, acetylglucosaminidase, and chitinase (Venugopal 1995). Enzymes which are more common to fish and aquatic invertebrates have also been documented in the literature. These include urease from fish liver (Kinsella et al. 1985), thymidilate kinase from sea urchin (Terentyev et al. 1999), and trimethylamine oxide demethylase from different fish species (Lundstrom et al. 1982).

The wide array of attainable enzymes and the diversity of tissues where these are found entail the existence of different extraction and isolation techniques. However, for digestive proteases, which is the most important group, many studies have reported the use of common and simple chromatographic procedures for various fish species such as tambaqui (*Colossoma macropomum*) (Marcuschi et al. 2010), Pacific cod (*Gadus macrocephalus*) (Fuchise et al. 2009), and walleye pollock (*Theragra chalcogramma*) (Kishimura et al. 2008). These protocols proved to be efficient in purifying fish proteases in a few steps and are relatively less expensive, being easily adapted to industrial scale. For instance, an alkaline peptidase (acting on polypeptides) was purified from the viscera of the silver mojarra (*Diapterus rhombeus*) in a three-step process involving heat treatment, ammonium sulphate fractionation, and molecular exclusion chromatography (Silva et al. 2011). Indeed, many studies have proposed the use of common and simple chromatographic procedures in the purification of gastric and intestinal enzymes from various fish species, such as the Nile tilapia (*Oreochromis niloticus*) (Bezerra et al. 2005) or skipjack tuna (*Katsuwonus pelamis*) (Klomklao et al. 2009). These protocols proved to be efficient in purifying fish trypsins in a few steps, and were relatively cheaper, being easily adapted to industrial scale and affording between 1 and 3 g of purified trypsin per 1 kg of wet waste. For instance, two kinds of trypsins were attained from the intestine of the skipjack tuna (Klomklao et al. 2009). The intestines were cut, homogenized into powder, and the homogenate was filtrated. The suspended intestine powder was centrifuged to remove tissue debris, being the supernatant lyophilized. This extract was purified by chromatography in a Sephacryl S-200 column. Further purification was done by dialysis and other chromatographic operations (Klomklao et al. 2009). Regarding enzymes other than proteases, hyaluronidase, for instance, was partially purified by acetone fractionation, ion-exchange column chromatography on a Type-I polystyrene-based anion-exchange resin,

Amberlite IRA 420, and subsequently by gel filtration on Sephadex G-200 (Krishnapillai et al. 1999). Hence, the association of various chromatographic steps may be required.

The main enzymes of fish and aquatic invertebrates and their by-products that may be potentially recovered for practical uses are presented in Table 1.

Table 1 Sources of enzymes from fish and aquatic invertebrates and their by-products (adapted from: Shahidi and Kamil 2001).

Group	Enzyme	Source
Gastric protease (aspartic protease family)	Pepsin	American smelt, capelin, mackerel, etc.
	Pepsinogen	Rainbow trout
	Gastricsin	Atlantic cod, hake
Intestinal proteinase (serine protease family)	Trypsin	Capelin, carp, krill, palometa, etc.
	Chymotrypsin	Capelin, grass carp, herring, etc.
	Collagenase	Atlantic cod, crayfish, etc.
	Elastase	Atlantic cod, carp, catfish
Chitinolytic enzyme	Chitinase (endo-β-N-acetylglucosaminidase)	Crustaceans, squid
	Lysozyme	Arctic scallop
Transglutaminases		Atka mackerel, walleye pollock liver, etc.
Lipases		Atlantic cod
Alkaline phosphatases		Arctic shrimp
Hyaluronidase		Norway lobster hepatopancreas

Even though no reliable market data are currently available, recovery of marine enzymes from underutilized fish resources is potentially profitable. Indeed, some established commercial applications already exist, such as fish curing and fermentation, but there are major opportunities for developing new niche markets (Ferraro et al. 2010). There is an outstanding profitable market for cod transglutaminase in surimi manufacture (Shahidi and Kamil 2001) as well as for lipases in the detergent industry (Esposito et al. 2009). Moreover, fish processing waste can be a source of alkaline proteases in the manufacture of laundry detergent (Esposito et al. 2009). Even though fish-derived enzymes do not have direct applications in the field of functional foods or nutraceuticals they can be utilized to produce bioactive components in large scale. Research targeting the preparation of crude enzyme mixtures from internal organs (by-products) of some fish species and utilization of them in the isolation of bioactive components from fish protein hydrolysates (Kim et al. 2003) has been reported recently. In addition, chitinases as well as chitosanases can also be isolated from the digestive tract and other organs of some marine fish species. These enzymes promote the recovery of chitin and chitosan from marine by-products necessary for a wide array of biomedical applications (Kim and Mendis 2006). A great number of recent research studies are focusing on novel purification methods and immobilization methodologies that can improve both stability and enantioselectivity of the enzymes and thus render enzymatic processes more competitive (Mohapatra et al. 2003). This can open new perspectives for the application of these enzymes.

PEPTIDES AND RELATED COMPOUNDS

Peptides with biological activities (e.g., antihypertensive, antioxidative) have a potential use as nutraceuticals and functional food additives. Marine fish viscera contain many useful proteins, which can be converted into value-added products by enzymatic hydrolysis (Kumar et al. 2011). Conversion of this underutilized raw material into useful products has environmental, economical, and biomedical applications. Bioactive peptides isolated from various fish protein hydrolysates have shown numerous bioactivities, namely, antihypertensive, antithrombotic, immunomodulatory, and antioxidative (Kim and Mendis 2006). Although the structure–activity relationship of many bioactive peptides has not yet been fully established, several structural features have been identified which appear to influence the biological action of peptides. For instance, regarding peptides that inhibit angiotensin converting enzyme (ACE (EC 3.4.15.1)), binding of the peptide to ACE is strongly influenced by the presence of tyrosine, phenylalanine, tryptophan, proline, lysine, isoleucine, valine, leucine, and arginine (López-Fandiño et al. 2006, Murray and FitzGerald 2007). Positively charged residues are associated with the activity of antimicrobial peptides (López Expósito and Recio 2006), whereas amino acid residues such as histidine, leucine, tyrosine, methionine, and cysteine are associated with radical scavenging activity. Hydrophobic amino acids such as proline and hydroxyproline seem to participate in the inhibition of lipid peroxidation (Mendis et al. 2005, Sarmadi and Ismail 2010). Among this diverse array of bioactivities, it has been reported that some fish-derived peptides present stronger antihypertensive activity than that of many other natural peptides (Kim et al. 2000). Moreover, a peptide from Alaska pollock (*Theragra chalcogramma*) frame protein hydrolysate exhibited *in vivo* activity by lowering blood pressure in spontaneously hypertensive rats (Je et al. 2005). Enzymatically hydrolyzed fish muscle peptides also have shown anticoagulant and antiplatelet properties tested *in vitro* (Rajapakse et al. 2005). The antioxidant properties of protein hydrolysate from sardinella (*Sardinella aurita*) viscera have also attracted wide attention of researchers (Nabil et al. 2007).

The type of bioactive peptides generated from a particular protein is depends on the primary sequence of the protein substrate and the enzymatic specificity associated with the generation of such peptides (Harnedy and Fitzgerald 2012). Different peptides can be generated by both acid and alkaline hydrolyses although this approach is generally not compatible with food additive manufacturing strategies. Hence, mild hydrolysis conditions with proteolytic enzymes are preferred in many cases. In general, a range of different processes are used to isolate and purify bioactive peptides from marine protein hydrolysates. Membrane separation techniques such as nanofiltration, ultrafiltration, and electro-membrane filtration provide a potentially suitable industrially-relevant technology for the enrichment of marine peptides within specific molecular weight ranges (Bourseau et al. 2009, Picot et al. 2010). Marine-derived bioactive peptides are usually fractioned using membranes in the range of 1–10 kDa (Hai-Lun et al. 2006, Je et al. 2005). In addition, membrane bioreactor technology which combines enzymatic hydrolysis of marine proteins and peptide separation by ultrafiltration is being considered as a potential method for the generation and fractionation of marine-derived bioactive peptides (Kim and Wijesekara 2010).

Particularly, for the purification of peptides with specific biofunctional activities, *in vitro* bioactivity assay-directed purification is used. But, in order to purify specific peptides, membrane processes may need to be associated with chromatographic techniques (Pouliot et al. 2006). On an industrial scale the processing costs associated with purifying a bioactive peptide must be assessed against the value of the purified or semi-purified product. Therefore, only semi- and preparative-scale chromatography (Fig. 6) may be required when high purity of the active peptides is crucial for commercialization (Pouliot et al. 2006). It must be noted that the extent of purification required will be dictated by the potency of the bioactive component. Besides, for any co-product stream arising from the purification process, outlets need to be found (Harnedy and Fitzgerald 2012).

Fish by-products and low-value fish species also contain a variety of biologically active amino acids. Taurine, a naturally occurring β-aminosulphonic acid, which is essential for numerous biological and physiological functions such as bile acid conjugation, cell membrane stabilization, development of the central nervous system and the retina, osmoregulation, modulation of cellular calcium levels, and immune function (Shao and Hathcock 2008). Although taurine has only recently become increasingly popular as an additive in dietary supplements,

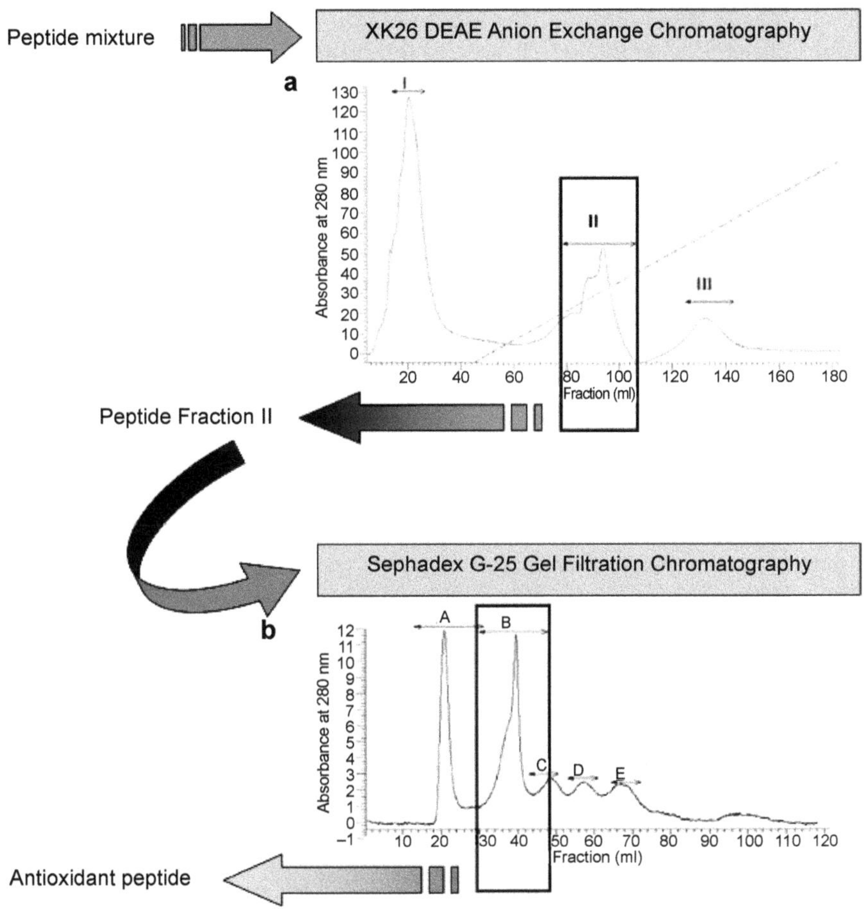

Fig. 6 Chromatographic purification of an antioxidant peptide from horse mackerel viscera protein (adapted from: Kumar et al. 2011).

functional foods, and beverages, it has been used in infant formulae since the 1980s (Harnedy and Fitzgerald 2012). It has been reported that extracts and hydrolysates of Alaska pollock and southern blue whiting (*Micromesistius australis*) contain 15 and 149 mg of taurine/100 g dry weight, respectively (Nakajima et al. 2009). Furthermore, the γ-aminobutyric acid (GABA), an amino acid like molecule synthesised from glutamate by two glutamic acid decarboxylase enzymes, has been found in all central nervous systems and represents the most common inhibitory neurotransmitter in vertebrates. Recently, GABA has been reported to reduce blood pressure in spontaneously hypertensive rats and in mildly hypertensive human subjects (Tsai et al. 2006). In addition to taurine, GABA has been observed as one of the most abundant components of the intracellular amino acid pool in the blood of some fish species (Fugelli 1970).

CHITIN, CHITOSAN AND RELATED PRODUCTS

Among the bioactive macromolecules that can be attained from marine by-products, chitin and its derivative chitosan are of great importance. These polysaccharides are linear copolymers of D-GlcN and D-GlcNAc residues distributed randomly and not blocked together, the residues are linked entirely in the β-1,4-configuration. This configuration leads to a rigid and unbranched structure (Prashanth and Tharanathan 2007). The abundance of hydroxyl groups and highly reactive amino groups or N-acetyl counterparts (wholly in chitin) with their propensity for intra- and intermolecular hydrogen bonds results in the formation of linear aggregates with extensive crystallinity. The latter contributes to the strength shown by chitinous structures.

Chitosan is commercially obtained mainly from chitin isolated from the shell wastes of crabs, shrimp, and krill (Kurita 2006) (Fig. 7). In fact, chitin is used almost exclusively as a raw material for the production of chitosan, oligosaccharides, and glucosamine, to levels that total worldwide 37,300 tons yearly (Chang et al. 2007). Shrimp cuticle has the highest amount of chitin, 30–40%, followed by crab's, 15–30% (Qin and Agboh 1997).

The extraction process at the industrial level involves treatment of the shells with dilute hydrochloric acid at room temperature in order to remove metals and salts such as calcium carbonate, followed by exposure to an alkaline medium to hydrolyse the proteins and pigments, and drying, thereby yielding a powdered product (Kurita 2006). Demineralization and deproteination of shrimp waste can also be carried out through enzymatic fermentation, which is a milder process than addition of hydrochloric acid (Kurita 2006, Sini et al. 2007). Proteases derived from *Bacillus subtilis* are used for chitin extraction from shrimp shells (Sini et al. 2007), whereas *Pseudomonas maltophilia* seems to be particularly effective in removing proteins from decalcified shrimp cuticle chips (Qin and Agboh 1997).

These polysaccharides have been the subject of various studies concerning their applications in the industry, food, pharmaceuticals and health care, as a result of their distinctive biological and physicochemical characteristics (Kurita 2006, Ferraro et al. 2010) (Table 2). Moreover, the current trend towards the use of natural products has contributed to the observed increase in their demand. However, high viscosity and low solubility at neutral pH may limit the application of these materials in functional food and pharmaceutical industries (Kim and Mendis 2006). Therefore currently, there is a growing interest to convert chitin and chitosan into their oligomers that have better functional properties. Production

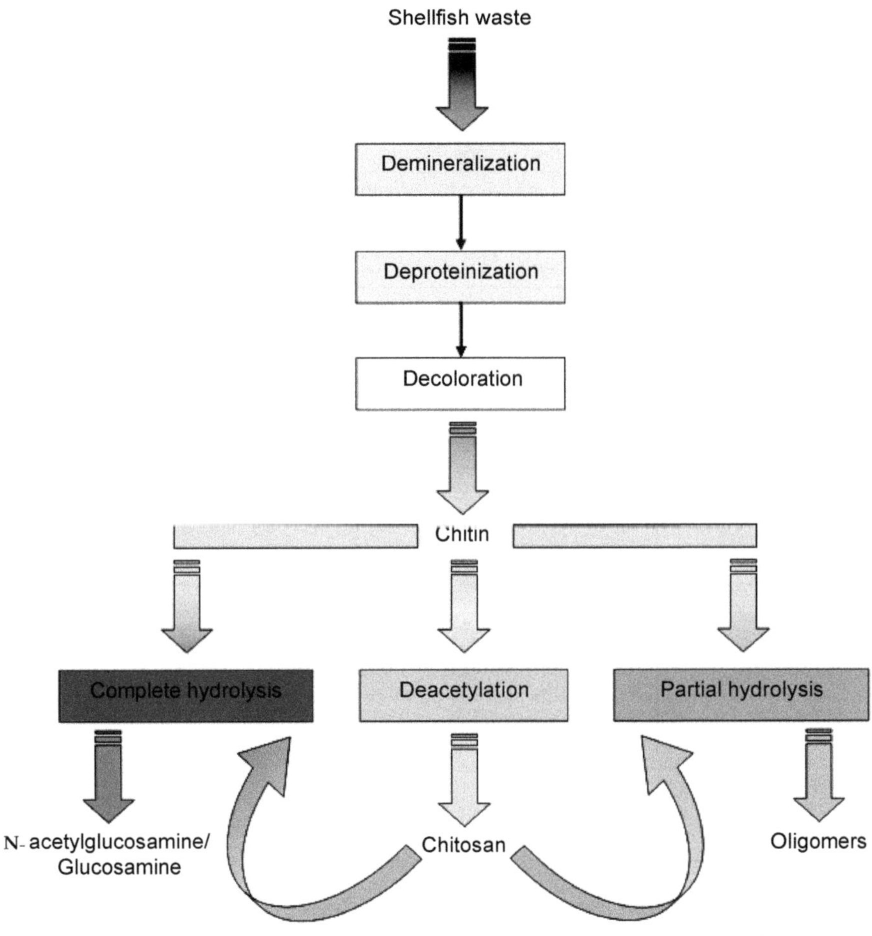

Fig. 7 Simplified flowsheet for preparation of chitin, chitosan, their oligomers and monomers from shellfish waste (adapted from: Shahidi et al. 1999).

of chitin and chitosan oligomers from hydrolysis can be done either by chemical or enzymatic methods. Chemical methods involve the use of acid hydrolysis and produce considerable amount of harmful industrial chemicals. On the other hand, enzymatic hydrolysis does not present this problem and leads to oligomers with higher degree of polymerization (Jeon and Kim 2000).

One of the major reasons for the use of chitosan as a functional additive is its hypocholesterolemic activity (Prashanth and Tharanathan 2007). This chitosan action has been explained on the basis of decreased cholesterol absorption and interference with bile acid absorption. Moreover, chitosan and several of its derivatives display antioxidant activity, given

Table 2 Applications of chitin and its derivatives (from: Rosa and Nunes 2008).

Application area	**Examples**
Foods	- Wine and juice clarifying agent - Dietary fibre, edible films - Protein flocculation, tannin removal
Agriculture	- Films for delaying fruit ripening - Seed coating - Controlled release of nutrients - Animal feeds
Water treatment	- Treatment of drinking water - Wastewater treatment, removal of metals, pesticides, and PCBs
Additives	- Oxidation inhibition - Thickening and stabilization - Texture modulation - Matrix for the slow release of additives
Biomedical	- Hypocholesterolemic effect - Wound treatment - Dental medicine - Drug transport
Cosmetics	- Hydration lotions, skin and hair treatment - Thickening in low pH products, emulsifying agent
Biotechnology	- Enzymatic and cellular immobilization - Nutraceutical encapsulation - Chromatography, analytical reagents - Protein recovery
Textile/paper	- Coatings, fibres - Coloration - Binding matrix

their protection from free radicals (Tiwari 2004). This antioxidant effect depends on chitosan molecular weight and viscosity, as shown in cooked minced fish model systems (Kamil et al. 2002). Indeed, it was shown that not only chitosan-treated herring samples had lower peroxide values and total volatile aldehydes than untreated samples, but also the low-viscosity chitosan presented the strongest antioxidative effect (Lin and Chou 2004). Chitosan oligomers also show scavenging activity on hydroxyl and superoxide radicals (Park et al. 2003). This property makes them not only potential additives for lipid oxidation inhibition in food, but also deterrents of pathological processes associated with free radical modification of cellular compounds, such as atherosclerosis, arthritis, diabetes, inflammatory disorders, and neurological disorders such as Alzheimer's disease (Blanco et al. 2007, Frölich and Riederer 1995). In addition, chitosan and its derivatives have shown antimicrobial activity. This results from the cationic charge of chitosan molecule, which favours aggressive binding onto the microbial cell surface, leading to gradual shrinkage of cell membrane and finally death of the cell. Chitosan has proven to be a broad-spectrum antimicrobial agent against both gram-positive and gram-negative bacteria and fungi (Kumar et al. 2005). It has also found orthopaedic medical applications (Prashanth and Tharanathan 2007). It has been reported that artificial bone formation using tissue-engineering technology is possible and that chitosan-tricalcium phosphate sponges may function as scaffold material for transplantation into a site for bone regeneration *in vivo*. The results of a preliminary *in vitro* experiment indicate that chitosan may foster the differentiation of osteoprogenitor cells and facilitate bone formation (Lee et al. 2000).

REGULATORY ASPECTS ASSOCIATED WITH THE USE OF MARINE NUTRACEUTICALS

Food regulation is aimed at protecting the consumer's health, increasing economic viability, harmonizing well-being, and engendering fair trade on foods within and between nations. For nutraceutical industries, two challenges are apparent, regulatory uncertainty and credibility of labelling claims (Aruoma 2006).

Nutraceuticals is a term coined (DeFelice 1995) to describe compounds that are not traditionally recognized nutrients, such as, vitamins and minerals, but which have positive physiological effects on the human body. Much of the early development of the nutraceutical concept and products was driven from the USA where the Dietary Supplement and Health Education Act has allowed considerable flexibility and blurred the boundaries between foods and medicines found in other parts of the world. However, the European Union (EU) has taken a less liberal stance on this matter.

Within EU law, the legal categorization of a nutraceutical is, in general, made on the basis of its accepted effects on the body. Thus, if the compound contributes only to the maintenance of healthy tissues and organs it is considered a food ingredient. If, however, it can be shown to have a modifying effect on one or more of the physiological processes, it is likely to be considered a medicinal substance (Gulati 2006). Under European law (Directive 2001/83/EC 2001, amended by Directive 2004/27/EC 2004) a nutraceutical can be defined as a medicine if it is presented for the prevention, treatment or cure of a condition or disease or if it can be administered with a view to restoring, correcting or modifying physiological functions in human beings (Gulati 2006).

When considering the original definition of a nutraceutical in terms of European law there is a high probability that many substances will fall into the classification of medicine rather than as a food ingredient. In the case of a nutraceutical being classified as food or food ingredient, Article 2 of Regulation (EC) No. 178/2002 (2002) giving the definition of a food ingredient must be taken into account. For whole foods with benefits beyond basic nutrition, the material must be well identified and characterized.

Food supplements are defined in Article 2 of Directive 2002/46/EC (2002), as "Food stuffs the purpose of which is to supplement the normal diet and which are concentrated sources of nutrients or other substances with a nutritional or physiological effect, alone or in combination, marketed in dose form namely forms such as capsules, pastilles, tablets, pills and other similar forms, sachets of powder, ampoules of liquids, drop dispensing bottles and other similar forms of liquids and powders designated to be taken in measured small unit quantities". This can be exemplified by ω3-PUFA concentrates.

Regarding labelling claims, the Regulation (EC) No. 1924/2006 (2006) establishes rules in labelling, presentation, and the advertising of foods and the Regulation (EC) No. 353/2008 (2008) implements rules for health claims in the Regulation (EC) No. 1924/2006 (2006). The important objective for the development of European legislation on this subject is to ensure that claims for food components and nutraceuticals are properly justified and they are scientifically substantiated (Byrne 2003). The mentioned EU legal framework enables the use of "health claims" and "reduction of disease risk claims" for foods outside the scope of medicinal law. Health claims are only approved for use on labelling, presentation, and advertising of foods in the EU market after a scientific evaluation of the highest possible standard. In order to ensure harmonized scientific assessment of a health claim, the European Food Safety Authority has a fundamental role in the performance of the assessments.

CONCLUSIONS

In order to attain a better utilization of fish waste and by-products and low-value fish species, food and health applications can and must be considered. Two main approaches can be adopted, either the preparation of restructured fish products (gelled or not) or extraction and isolation of valuable bioactive molecules. Of course, these two strategies can be combined by means of production of restructured fish products enriched in biomolecules, such as fish sausages from hake trimmings enriched with ω3-PUFA from fish by-products.

Among the bioactive molecules, besides lipids such as ω3-PUFA, enzymes (namely proteases), peptides, and chitin/chitosan and related compounds are of outstanding importance. These molecules are not only addable to foods as nutraceutical ingredients, but can exert a clear pharmaceutical effect, for instance, in the case of peptides, as antihypertensive, antithrombotic, and immunomodulatory agents. These applications demand purification technologies, which may lead to lengthy processes with many and sophisticated operations, such as preparative chromatography. Though the recovery of valuable molecules from seafood by-products is a challenging task for the scientists, the commercial implementation is also a difficulty, which depends on several parameters. Researchers must manage to scale up without affecting the functional properties of the target compound and to develop a product with high safety and quality standards. This task is not as simple in practice as it seems. For instance, problems due to the season effect and raw material variety and composition have to be solved. This affects the character or functionality of the final product. Hence, such problems can only be solved in terms of an integrated approach to resource management through utilization of the most recent technological advances. A commercially feasible product requires that developed methodology presents a leeway for a certain degree of flexibility and alternative choices in the raw materials. Concerning this issue, simple processes not only tend to be easier to scale up, but are also economically advantageous.

On the other hand, the rapid technological changes in this field of research present a challenge to the legislator, which has recently put forward significant legislation, namely, concerning the scientific soundness of health claims. Nevertheless, drawing a dividing line between food and health applications remains difficult, thereby posing a quandary for legislators, regulators, industrialists, and scientists.

Finally, the commitment of all actors in the fisheries sector is an important and indispensable requirement underlying the successful solution of all mentioned challenges.

ACKNOWLEDGMENTS

This research was funded by SECUREFISH project FP 7 EU PROJECT THEME [KBBE.2011.2.5-02] Grant agreement No 289282 "Improving Food Security by Reducing Post Harvest Losses in the Fisheries Sector".

REFERENCES

Aidos, I., A. van-der-Padt, R.M. Boom and J.B. Luten. 2001. Upgrading of maatjes herring byproducts: Production of crude fish oil. J. Agric. Food Chem. 49: 3697–3704.

Aidos, I., A. van-der-Padt, R.M. Boom and J.B. Luten. 2003a. Quality of crude fish oil extracted from herring by-products of varying states of freshness. J. Food Sci. 68: 458–465.

Aidos, I.,N. Kreb, M. Boonman, J.B. Luten, R.M. Boom and A. van-der-Padt. 2003b. Influence of production process parameters on fish oil quality in a pilot plant. J. Food Sci. 68: 581–587.

Aruoma, O.I. 2006. The impact of food regulation on the food supply chain. Toxicol. 221: 119–127.
Babbit, J.K. and K.D. Reppond. 1987. Manufacturing of a crab analogue to determine the quality of U.S. shore-based produced surimi. Mar. Fisheries Rev. 49(4): 34–36.
Batista, I. and C. Pires. 2002. Comparative studies of the proteolytic activity of crude extracts from the digestive tract of three shark species. J. Aq. Food Prod. Technol. 11: 151–165.
Bentis, C.A., A. Zotos and D. Petridis. 2005. Production of fish-protein products (surimi) from small pelagic fish (Sardinops pilchardus), underutilized by the industry. J. Food Eng. 68: 303–308.
Bertak, J.A. and C. Karahadian. 1995. Surimi-based imitation crab characteristics affected by heating method and end point temperature. J. Food Sci. 60: 292–296.
Beynon, R.J. and J.S. Bond. 2001. Proteolytic enzymes: A practical approach. *In:* J.N. Michael and R.J. Beynon [eds.]. Prevention of unwanted proteolysis. Oxford University Press, Oxford, UK pp. 211–231.
Bezerra, R.S., E.J.F. Lins, R.B. Alencar, P.M.G. Paiva, M.E.C. Chaves, L.C.B. Coelho and L.B. Carvalho Jr. 2005. Alkaline proteinase from intestine of Nile tilapia (Oreochromis niloticus). Proc. Biochem. 40: 1829–1834.
Blanco, M., C.G. Sotelo, M.J. Chapela and R.I. Pérez-Martín. 2007. Towards sustainable and efficient use of fishery resources: present and future trends. Trends Food Sci. Technol. 18: 29–36.
Bochi, V.C., J. Weber, C.P. Ribeiro, A.M. Victório and T. Emanuelli. 2008. Fishburgers with silver catfish (Rhamdia quelen) filleting residue. Biores. Technol. 99: 8844–8849.
Borderías, A.J., I. Sánchez-Alonso and M. Pérez-Mateos. 2005. New applications of fibres in foods: addition to fishery products. Trends Food Sci. Technol. 16: 458–465.
Bourseau, P., L. Vandanjon, P. Jaouen, M. Chaplain-Derouiniot, A. Massé, F. Guérard, A. Chabeaud, M. Fouchereau-Péron, Y. Le Gal, R. Ravallec-Plé, J.P. Bergé, L. Picot, J.M. Piot, I. Batista, G. Thorkelsson, C. Delannoy, G. Jakobsen and I. Johansson. 2009. Fractionation of fish protein hydrolysates by ultrafiltration and nanofiltration: Impact on peptidic populations. Desalination 244: 303–320.
Byrne, D. 2003. Health nutrition and labeling. Food Sci. Technol. 17: 26–28.
Cardoso, C., R. Mendes and M.L. Nunes. 2007a. Effect of transglutaminase and carrageenan on restructured fish products containing dietary fibres. Int. J. Food Sci. Technol. 42(11): 1257–1264.
Cardoso, C., R. Mendes and M.L. Nunes. 2007b. Dietary fibers' effect on the textural properties of fish heat-induced gels. J. Aq. Food Prod. Technol. 16(3): 19–30.
Cardoso, C., R. Mendes and M.L. Nunes. 2008a. Development of a healthy low fat fish sausage containing dietary fibre. Int. J. Food Sci. Technol. 43(2): 276–283.
Cardoso, C., R. Mendes, S. Pedro and M.L. Nunes. 2008b. Quality changes during storage of fish sausages containing dietary fibre. J. Aq. Food Prod. Technol. 17(1): 73–95.
Cardoso, C., R. Mendes and M.L. Nunes. 2009a. Instrumental texture and sensory characteristics of cod Frankfurter sausages. Int. J. Food Prop. 12(3): 625–643.
Cardoso, C., R. Mendes, P. Vaz-Pires and M.L. Nunes. 2009b. Effect of dietary fibre and MTGase on the quality of mackerel surimi gels. J. Sci. Food Agric. 89(10): 1648–1658.
Cardoso, C., R. Mendes, S. Pedro, P. Vaz-Pires and M.L. Nunes. 2010. Quality changes during storage of minced fish products containing dietary fibre and fortified with Ω 3 fatty acids. Food Sci. Technol. Int. 16(1): 0031–42.
Cardoso, C., R. Mendes, P. Vaz-Pires and M.L. Nunes. 2011. Production of high quality gels from sea bass: effect of MTGase and dietary fibre. Food Sci. Technol. 44(5): 1282–1290.
Cardoso, C., B. Ribeiro and R. Mendes. 2012. Improvement of the gelling ability in restructured fish products: Effect of 1-ethyl-3-(3-dimethylaminopropyl) carbodiimide level and pH. Eur. Food Res. Technol. 234(6): 935–943.
Chang, W.T., Y.C. Chen and C.L. Jao. 2007. Antifungal activity and enhancement of plant growth by Bacillus cereus grown on shellfish chitin wastes. Biores. Technol. 98: 1224–1230.

Chantachum, S., S. Benjakul and N. Sriwirat. 2000. Separation and quality of fish oil from precooked and non-precooked tuna heads. Food Chem. 69: 289–294.
Chen, H.H., E.M. Chiu and J.R. Huang. 1997. Colour and gel forming properties of horse mackerel (Trachurus japonicus) as related to washing conditions. J. Food Sci. 62: 985–991.
Chen, Y.C. and J. Jaczynski. 2007. Protein recovery from rainbow trout (Oncorhynchus mykiss) processing by-products via isoelectric solubilization/precipitation and its gelation properties as affected by functional additives. J. Agric. Food Chem. 55(22): 9079–9088.
Chen, Y.C., J.C. Tou and J. Jaczynski. 2007. Amino acid, fatty acid, and mineral profiles of materials recovered from rainbow trout (Oncorhynchus mykiss) processing by-products using isoelectric solubilization/precipitation. J. Food Sci. 72(9): C527–C535.
Chen, Y.C., J.C. Tou and J. Jaczynski. 2009. Amino acid and mineral composition of protein and other components and their recovery yields from whole Antarctic krill (Euphausia superba) using isoelectric solubilization/precipitation. J. Food Sci. 74(2): H31–H39.
Daukšas, E., E. Falch, R. Šližyte and T. Rustad. 2005. Composition of fatty acids and lipid classes in bulk products generated during enzymic hydrolysis of cod (Gadus morhua) by-products. Proc. Biochem. 40: 2659–2670.
DeFelice, S.L. 1995. The nutritional revolution: Its impact on food industry R&D. Trends Food Sci. Technol. 6: 59–61.
Directive 2001/83/EC of the European Parliament and of the Council 2001. Community code relating to medicinal products for human use. Off. J. EU. L-311: 0067–0128.
Directive 2002/46/EC of the European Parliament and of the Council 2002. Approximation of the laws of member states relating to food supplements. Off. J. EU. L-183: 0051–0057.
Directive 2004/27/EC of the European Parliament and of the Council 2004. Amending Directive 2001/83/EC of the European Parliament and of the Council of 6 November 2001 on the community code relating to medicinal products for human use. Off. J. EU. 136: 0034–0057.
Draves, R. 2003. Surimi processing. Presented at the 11th OSU Surimi School, Astoria, USA.
Espósito, T.S., I.P.G. Amaral, D.S. Buarque, G.B. Oliveira, L.B. Carvalho Jr. and R.S. Bezerra. 2009. Fish processing waste as a source of alkaline proteases for laundry detergent. Food Chem. 112: 125–130.
FAO, F.I.D. 1986. The production of fish meal and oil. F.F.T. Paper, vol. 142, FAO, Rome, Italy. 63 pp.
Ferraro, V., I.B. Cruz, R.F. Jorge, F.X. Malcata, M.E. Pintado and P.M.L. Castro. 2010. Valorisation of natural extracts from marine source focused on marine by-products: A review. Food Res. Int. 43: 2221–2233.
Frölich, I. and P. Riederer. 1995. Free radical mechanisms in dementia of Alzheimer type and the potential for antioxidative treatment. Drug Resistance Updates 45: 443–446.
Fuchise, T., H. Kishimura, H. Sekizaki, Y. Nonami, G. Kanno, S. Klomklao, S. Benjakul and B.S. Chun. 2009. Purification and characteristics of trypsins from cold-zone fish, Pacific cod (Gadus macrocephalus) and saffron cod (Eleginus gracilis). Food Chem. 116(3): 611–616.
Fugelli, K. 1970. Gamma-aminobutyric acid (GABA) in fish erythrocytes. Cell. Mol. Life Sci. 26: 361.
Gehring, C.K., J.C. Gigliotti, J.S. Moritz, J.C. Tou and J. Jaczynski. 2011. Functional and nutritional characteristics of proteins and lipids recovered by isoelectric processing of fish by-products and low-value fish: A review. Food Chem. 124: 422–431.
Gómez-Guillén, M.C., T. Solas and P. Montero. 1997. Influence of added salt and non-muscle proteins on the rheology and ultrastructure of gels made of minced flash of sardine (Sardina pilchardus). Food Chem. 58: 193–202.
Guenneugues, P. and M.T. Morrissey. 2005. Surimi resources. *In:* J.W. Park [ed.]. Surimi and surimi seafood. CRC press, Boca Raton, USA pp. 3–32.
Gulati, O.P. and P.B. Ottaway. 2006. Legislation relating to nutraceuticals in the European Union with a particular focus on botanical-sourced products. Toxicol. 221: 75–87.

Hai-Lun, H., C. Xiu-La, S. Cai-Yun, Z. Yu-Zhong and Z. Bai-Cheng. 2006. Analysis of novel angiotensin-I-converting enzyme inhibitory peptides from protease-hydrolyzed marine shrimp Acetes chinensis. J. Peptide Sci. 12: 726–733.
Harnedy, P.A. and R.J. Fitzgerald. 2012. Bioactive peptides from marine processing waste and shellfish: A review. J. Funct. Foods 4: 6–24.
Hollingworth, T.A., C.A. Kaysner, K.G. Colburn, J.J. Sullivan, C. Abeuta Jr., K.D. Walker, J.D. Torkelson, H.R. Throm and M.M. Wekell. 1991. Chemical and microbiological analysis of vacuum-packed, pasteurized flaked imitation crabmeat. J. Food Sci. 56: 164–167.
Hultin, H.O. and S.D. Kelleher. 1999. Process for isolating a protein composition from a muscle source and protein composition. U.S. Patent No. 6,005,073. Granted on 21st December 1999.
Hultin, H.O. and S.D. Kelleher. 2000. High efficiency alkaline protein extraction. U.S. Patent No. 6,136,959. Granted on 24th October 2000.
Hultin, H.O., H.G. Kristinsson, T.C. Lanier and J.W. Park. 2005. Process for recovery of functional proteins by pH shifts. *In:* J.W. Park [ed.]. Surimi and surimi seafood. CRC press, Boca Raton, USA pp. 107–139.
Je, J.Y., S.Y. Kim and S.K. Kim. 2005. Preparation and antioxidative activity of hoki frame protein hydrolysate using ultrafiltration membranes. Eur. Food Res. Technol. 221: 157–162.
Je, J.Y., P.J. Park, W.K. Jung and S.K. Kim. 2005. Isolation of angiotensin I converting enzyme (ACE) inhibitor from fermented oyster sauce, Crassostrea gigas. Food Chem. 90: 809–814.
Je, J.Y., P.J. Park, J.Y. Kwon and S.K. Kim. 2005. A novel angiotensin I converting enzyme inhibitory peptide from Allaska Pollack (Theragra chalcogramma) frame protein hydrolysate. J. Agric. Food Chem. 52: 7842–7845.
Jeon, Y.J. and S.K. Kim. 2000. Continuous production of chitooligosaccharides using a dual reactor system. Proc. Biochem. 35: 623–632.
Jeon, Y.J. and S.K. Kim. 2002. Antitumor activity of chitosan oligosaccharides produced in an ultra filtration membrane reactor system. J. Microbiol. Biotechnol. 12: 503–507.
Kamil, J., Y.J. Jeon and F. Shahidi. 2002. Antioxidant activity of chitosans of different viscosity in cooked comminuted flesh of herring (Clupea harengus). Food Chem. 79: 69–77.
Kim, J.M., C.H. Liu, J.B. Eun, J.W. Park, R. Oshimi, K. Hayashi, B. Ott, T. Aramaki, M. Sekine, V. Horikita, K. Fujimoto, T. Alkawa, L. Welch and R. Long. 1996. Surimi from fillet frames of Channel catfish. J. Food Sci. 61: 428–431, 438.
Kim, J.S., J.W. Park. 2007. Mince from seafood processing by-product and surimi as food ingredients. *In:* F. Shahidi. [ed.]. Maximising the value of marine by-products. Woodhead Publishing Ltd., Cambridge, UK pp. 196–228.
Kim, S.K., Y.R. Choi, P.J. Park, J.H. Choi and S.H. Moon. 2000. Screening of biofunctional peptides from cod processing wastes. J. Kor. Soc. Agric. Chem. Biotechnol. 43: 225–227.
Kim, S.K., Y.T. Kim, H.G. Byun, K.S. Nam, D.S. Joo and F. Shahidi. 2001. Isolation and characterization of antioxidative peptides from gelatin hydrolysate of Allaska Pollack skin. J. Agric. Food Chem. 49: 1984–1989.
Kim, S.K., P.J. Park, H.G. Byun, J.Y. Je and S.H. Moon. 2003. Recovery of fish bone from hoki (Johnius belengeri) frame using a proteolytic enzyme isolated from mackerel intestine. J. Food Biochem. 27: 255–266.
Kim, S.K. and E. Mendis. 2006. Bioactive compounds from marine processing byproducts—A review. Food Res. Int. 39: 383–393.
Kim, S.K. and I. Wijesekara. 2010. Development and biological activities of marine-derived bioactive peptides: A review. J. Funct. Foods 2: 1–9.
Kinsella, J.E., J.B. German and J. Shetty. 1985. Urease from fish liver: isolation and some properties. Comp. Biochem. Physiol. 82B: 621–624.
Kishimura, H., S. Klomklao, S. Benjakul and B.S. Chun. 2008. Characteristics of trypsin from ceca of walleye pollock (Theragra chalcogramma). Food Chem. 106: 194–199.

Klomklao, S., H. Kishimura, Y. Nonami and S. Benjakul. 2009. Biochemical properties of two isoforms of trypsin purified from the intestine of skipjack tuna (Katsuwonus pelamis). Food Chem. 115: 155–162.

Kris-Etherton, P.M., W.S. Harris and L.J. Appel. 2002. American Heart Association, Nutrition Committee. Fish consumption, fish oil, omega-3 fatty acids, and cardiovascular disease. Circulation 106: 2747–2757.

Krishnapillai, A.M., K.D.A. Taylor, A.E.J. Morris and P.C. Quantick. 1999. Extraction and purification of hyaluronoglucosidase (EC 3.2.1.35) from Norway lobster (Nephrops norvegicus). Food Chem. 65: 359–365.

Kristinsson, H.G. and H.O. Hultin. 2003. Changes in conformation and subunit assembly of cod myosin at low and high pH and after subsequent refolding. J. Agric. Food Chem. 51(24): 7187–7196.

Kristinsson, H.G., A.E. Theodore, N. Demir and B. Ingadottir. 2005. A comparative study between acid- and alkali-aided processing and surimi processing for the recovery of proteins from channel catfish muscle. J. Food Sci. 70: C298–C306.

Kristinsson, H.G. and Y. Liang. 2006. Effect of pH-shift processing and surimi processing on Atlantic croaker (Micropogonias undulates) muscle proteins. J. Food Sci. 71: 304–312.

Kumar, A.B.V., M.C. Varadaraj, L.R. Gowda and R.N. Tharanathan. 2005. Characterization of chitooligosaccharides prepared by chitosanolysis with the aid of papain and pronase, and their bactericidal action. Biochem. J. 391: 167–175.

Kumar, N.S.S., R.A. Nazeer and R. Jaiganesh. 2011. Purification and biochemical characterization of antioxidant peptide from horse mackerel (Magalaspis cordyla) viscera protein. Peptides 32: 1496–1501.

Kuraishi, C., J. Sakamoto, K. Yamanazaki, Y. Susa, C. Kuhara and C. Soeda. 1997. Production of restructured meat using microbial transglutaminase without salt or cooking. J. Food Sci. 62(3): 488–490, 515.

Kurita, K. 2006. Chitin and chitosan: Functional biopolymers from marine crustaceans. Mar. Biotechnol. 8: 203–226.

Lanier, T.C. 1986. Functional properties of surimi. Food Technol. 40(3): 107–114.

Lanier, T.C. 1994. Functional food protein ingredients from fish. *In:* Z.E. Sikorski, B.S. Pan and F. Shahidi [eds.]. Seafood proteins. Chapman and Hall, Inc., New York, USA pp. 127–159.

Lanier, T.C., P. Carvajal and J. Yongsawatdigul. 2005. Surimi gelation chemistry. *In:* J.W. Park [ed.]. Surimi and surimi seafood. CRC press, Boca Raton, USA pp. 435–489.

Lee, Y.M., Y.J. Park, S.J. Lee, Y. Ku, S.B. Han, S.M. Choi, P.R. Klokkevold and C.P. Chung. 2000. Tissue engineered bone formation using chitosan/tricalcium phosphate sponges. J. Periodontol. 71: 410–417.

Liang, Y. and H.O. Hultin. 2005. Separation of muscle membrane from alkalisolubilized fish muscle proteins. J. Agric. Food Chem. 53: 10012–10017.

Liaset, B., K. Julshamn and M. Espe. 2003. Chemical composition and theoretical nutrition evaluation of the produced fractions from enzyme hydrolysis of salmon with Protamex™. Proc. Biochem. 38: 1747–1759.

Lin, H.Y. and C.C. Chou. 2004. Antioxidative activities of water soluble disaccharides of chitosan derivatives. Food Res. Int. 37: 883–889.

Linder, M., J. Fanni and M. Parmentier. 2005. Proteolytic extraction of salmon oil and PUFA concentration by lipases. Mar. Biotechnol. 7(1): 70–76.

López Expoósito, I. and I. Recio. 2006. Antibacterial activity of peptides and folding variants from milk proteins. Int. Dairy J. 16: 1294–1305.

López-Fandiño, R., J. Otte and J. Van Camp. 2006. Physiological, chemical and technological aspects of milk-protein-derived peptides with antihypertensive and ACE-inhibitory activity. Int. Dairy J. 16: 1277–1293.

Lundstrom, R.C., F.F. Correia and K.A. Wilhelm. 1982. Enzymatic dimetylamine and formaldehyde production in minced American plaice and blackback flounder mixed with red hake TMAO-ase active fraction. J. Food Sci. 47: 1305–1307.

Luo, Y., H. Shen, D. Pan and G. Bu. 2008. Gel properties of surimi from silver carp (Hypophthalmichthys molitrix) as affected by heat treatment and soy protein isolate. Food Hydrocolloids 22(8): 1513–1519.

Marcuschi, M., T.S. Espósito, M.F.M. Machado, I.Y. Hirata, M.F.M. Machado, M.V. Silva, L.B. Carvalho Jr., V. Oliveira and R.S. Bezerra. 2010. Purification, characterization and substrate specificity of a trypsin from the Amazonian fish tambaqui (Colossoma macropomum). Biochem. Biophys. Res. Comm. 396: 667–673.

Mendis, E., N. Rajapakse, H.G. Byun and S.K. Kim. 2005. Investigation of jumbo squid (Dosidicus gigas) skin gelatine peptides for their *in vitro* antioxidant effects. Life Sci. 77: 2166–2178.

Meyers, M.A. and K.K. Chawla. 1998. Mechanical behavior of materials. Prentice-Hall, New Jersey, USA.

Mohapatra, B.R., M. Bapuji and A. Sree. 2003. Production of industrial enzymes (amylase, carboxymethylcellulase and protease) by bacteria isolated from marine sedentary organizms. Acta Biotechnologica 23(1): 75–84.

Moreno, H.M., J. Carballo and A.J. Borderías. 2008. Influence of alginate and microbial transglutaminase as binding ingredients on restructured fish muscle processed at low temperature. J. Sci. Food Agric. 88: 1529–1536.

Murray, B.A. and R.J. FitzGerald. 2007. Angiotensin converting enzyme inhibitory peptides derived from food proteins: Biochemistry, bioactivity and production. Curr. Pharmac. Design 13: 773–791.

Nabil, S., B. Ali, Y. Triki-Ellouz and M. Nasri. 2007. Biochemical and functional properties of Sardinella (Sardinella aurita) by-product hydrolysates. Food Technol. Biotechnol. 45(2): 187–94.

Nakajima, K., Y. Yoshie-Stark and M. Ogushi. 2009. Comparison of ACE inhibitory and DPPH radical scavenging activities of fish muscle hydrolysates. Food Chem. 114: 844–851.

Nolsoe, H. and I. Undeland. 2009. The acid and alkaline solubilization process for the isolation of muscle proteins: State of the art. Food Bioprocess Technol. 2: 1–27.

Park, J.W. and T.M.J. Lin. 2005. Surimi: manufacturing and evaluation. *In:* J.W. Park [ed.]. Surimi and surimi seafood. CRC press, Boca Raton, USA pp. 33–106.

Park, P.J., J.Y. Je and S.K. Kim. 2003. Free radical scavenging activity of chitooligosaccharides by electron spin resonance spectrometry. J. Agric. Food Chem. 51: 4624–4627.

Picot, L., R. Ravallec, M. Fouchereau-Péron, L. Vandanjon, P. Jaouen, M. Chaplain-Derouiniot, F. Guérard, A. Chabeaud, Y. Legal, O. Martinez Alvarez, J.P. Bergé, J.M. Piot, I. Batista, C. Pires, G. Thorkelsson, C. Delannoy, G. Jakobsen, I. Johansson and P. Bourseau. 2010. Impact of ultrafiltration and nanofiltration of an industrial fish protein hydrolysate on its bioactive properties. J. Sci. Food Agric. 90: 1819–1826.

Pires, C., I. Batista, P. Fradinho and S. Costa. 2009. Utilization of alkaline-recovered proteins from Cape hake by-products in the preparation of Frankfurter-type fish sausages. J. Aq. Food Prod. Technol. 18: 170–190.

Pires, C., S. Costa, A.P. Batista, M.C. Nunes, A. Raymundo and I. Batista. 2012. Properties of protein powder prepared from Cape hake by-products. J. Food Eng. 108: 268–275.

Pouliot, Y., S.F. Gauthier and P.E. Groleau. 2006. Membrane-based fractionation and purification strategies for bioactive peptides. *In:* Y. Mine and F. Shahidi [eds.]. Nutraceutical proteins and peptides in health and disease. CRC Press, Boca Raton, USA pp. 639–658.

Prashanth, K.V.H. and R.N. Tharanathan. 2007. Chitin/chitosan: modifications and their unlimited application potential—an overview. Trends Food Sci. Technol. 18: 117–131.

Qin, Y. and O.C. Agboh. 1997. Chitin and chitosan fibers. Polymers Adv. Technologies 8: 355–365.

Raa, J. 1990. Biotechnology in aquaculture and the fish processing industry: a success story in Norway. *In:* M.N. Voigt and J.R. Botta [eds.]. Advances in fisheries technology for increased profitability. Technomic Publication Company, Lancaster, USA pp. 509–524.

Rajapakse, N., W.K. Jung, E. Mendis, S.H. Moon and S.K. Kim. 2005. A novel anticoagulant purified from fish protein hydrolysate inhibits factor XIIa and platelet aggregation. Life Sci. 76: 2607–2619.

Ramírez, J.A., N.R. Rodríguez, R.M. Uresti, G. Velazquez and M. Vázquez. 2007a. Fibre-rich functional fish food from striped mullet (Mugil cephalus) using amidated low methoxyl pectin. Food Hydrocolloids 21: 527–536.

Ramírez, J.A., A. Del Ángel, R.M. Uresti, G. Velásquez and M. Vázquez. 2007b. Low-salt restructured products from striped mullet (Mugil cephalus) using microbial transglutaminase or whey protein concentrate as additives. Food Chem. 102: 243–249.

Rasmussen, R.S. and M.T. Morrissey. 2007. Marine biotechnology for production of food ingredients. Adv. Food Nut. Res. 52: 237–292.

Rawdkuen, S., S. Sai-Ut, S. Khamsorn, M. Chaijan and S. Benjakul. 2009. Biochemical and gelling properties of tilapia surimi and protein recovered using an acid-alkaline process. Food Chem. 112: 112–119.

Reece, P. 1988. Recovery of proteases from fish waste. Proc. Biochem. 6: 62–66.

Regulation (EC) No. 178/2002 of the European Parliament and of the Council 2002. Laying down the general principles and requirements of food law, establishing the European Food Safety Authority and laying down procedures in matters of food safety. Off. J. EU. L 31: 0001–0024.

Regulation (EC) No 1924/2006 of the European Parliament and of the Council 2006. On nutrition and health claims made on foods. Off. J. EU. L 15: 244–259.

Regulation (EC) No 353/2008 of the European Parliament and of the Council 2008. Establishing implementing rules for applications for authorization of health claims as provided for in Article 15 of Regulation (EC) No 1924/2006 of the European Parliament and of the Council. Off. J. EU. L 109: 11–16.

Rosa, R. and M.L. Nunes. 2008. Crustáceos: Exploração, bioquímica, conservação e aproveitamento de sub-produtos. Publicações Avulsas do Instituto de Investigação das Pescas e do Mar, IPIMAR, Lisboa, Portugal.

Rubio-Rodríguez, N., S. Beltrán, I. Jaime, S.M. de Diego, M.T. Sanz and J.R. Carballido. 2010. Production of omega-3 polyunsaturated fatty acid concentrates: A review. Innov. Food Sci. Emerg. Technologies 11: 1–12.

Sarmadi, B.H. and A. Ismail. 2010. Antioxidative peptides from food proteins: A review. Peptides 31: 1949–1956.

Shahidi, F., J.K.V. Arachchi and Y.J. Jeon. 1999. Food applications of chitin and chitosans. Trends Food Sci. Technol. 10: 37–51.

Shahidi, F. and Y.V.A.J. Kamil. 2001. Enzymes from fish and aquatic invertebrates and their application in the food industry. Trends Food Sci. Technol. 12: 435–464.

Shao, A. and J.N. Hathcock. 2008. Risk assessment for the amino acids taurine, L-glutamine and L-arginine. Regul. Toxicol. Pharmacol. 50: 376–399.

Shimizu, Y., H. Toyohara and T.C. Lanier. 1992. Surimi production from fatty and dark flash species. *In:* T. Lanier and C. Lee [eds.]. Surimi technology. Marcel Dekker, Inc., New York, USA pp. 181–207.

Silva, J.F., T.S. Espósito, M. Marcuschi, K. Ribeiro, R.O. Cavalli, V. Oliveira and R.S. Bezerra. 2011. Purification and partial characterisation of a trypsin from the processing waste of the silver mojarra (Diapterus rhombeus). Food Chem. 129: 777–782.

Simopoulos, A.P. 2002. Omega-3 fatty acids and cardiovascular disease: the epidemiological evidence. Environ. Health Prev. Med. 6: 203–209.

Sini, K.T., S. Santosh and T.P. Mathew. 2007. Study on the production of chitin and chitosan from shrimp shell by using Bacillus subtilis fermentation. Carbohydrate Res. 342: 2423–2429.

Šližyte, R., E. Daukšas, E. Falch, I. Storrø and T. Rustad. 2005. Yield and composition of different fractions obtained after enzymatic hydrolysis of cod (Gadus morhua) byproducts. Proc. Biochem. 40: 1415–1424.

Tahergorabi, R., S.K. Beamer, K.E. Matak and J. Jaczynski. 2012. Functional food products made from fish protein isolate recovered with isoelectric solubilization/precipitation. LWT–Food Sci. Technol. 48(1): 89–95.

Taskaya, L., Y.C. Chen, S. Beamer and J. Jaczynski. 2009. Texture and colour properties of proteins recovered from whole gutted silver carp (Hypophthalmichthys molitrix) using isoelectric solubilization/precipitation. J. Sci. Food Agric. 89(2): 349–358.

Taskaya, L., Y.C. Chen and J. Jaczynski. 2010. Color improvement by titanium dioxide and its effect on gelation and texture of proteins recovered from whole fish using isoelectric solubilization/precipitation. LWT–Food Sci. Technol. 43(3): 401–408.

Taskaya, L. and J. Jaczynski. 2009. Flocculation-enhanced protein recovery from fish processing by-products by isoelectric solubilization/precipitation. LWT–Food Sci. Technol. 42(2): 570–575.

Téllez-Luis, S.J., R.M. Uresti, J.A. Ramírez and M. Vázquez. 2002. Low-salt restructured fish products using microbial transglutaminase as binding agent. J. Sci. Food Agric. 82: 953–959.

Terentyev, L.L., N.A. Terentyeva and V.A. Rasskazov. 1999. Purification and some properties of thymidilate kinase from sea urchin. Biochemistry (Moscow) 64: 80–85.

Tiwari, A.K. 2004. Antioxidants: new generation therapeutic base for treatment of polygenic disorders. Curr. Sci. 86: 1092–1102.

Torres, J.A., Y.C. Chen, J. Rodrigo-Garcia and J. Jaczynski. 2007. Recovery of byproducts from seafood processing streams. *In:* F. Shahidi [ed.]. Maximising the value of marine by-products. CRC Press, Boca Raton, USA pp. 65–90.

Tsai, J.S., Y.S. Lin, B.S. Pan and T.J. Chen. 2006. Antihypertensive peptides and [gamma]-aminobutyric acid from prozyme 6 facilitated lactic acid bacteria fermentation of soymilk. Proc. Biochem. 41: 1282–1288.

Venugopal, V. 1995. By-products from industrial fishery processing. Indian Food Ind. 14: 22–24.

Yoon, K.S., C.M. Lee and L.A. Hufnagel. 1991. Effect of washing on the texture and microstructure of frozen fish mince. J. Food Sci. 56: 294–298.

3

Functional Properties of Fish Protein Hydrolysates

*Carla Pires** and *Irineu Batista*

ABSTRACT

The recognized functional and biological properties of fish protein hydrolysates make them an alternative for the upgrading of by-products and underutilized species. The interest for these products is evidenced by the increasing number of recently published papers on this subject and also by the setting-up of several producing industries in different fishing countries. Among fish protein hydrolysates properties their functional properties are particularly important because they determine their utilization in food systems. This review describes the functional properties of these hydrolyzed proteins which include solubility, emulsification/emulsifying properties, fat absorption/fat binding capacity, foaming properties, water holding capacity/water absorption capacity, and sensory properties. The effect of manufacturing conditions on functional properties and possible correlations between the physicochemical peptide characteristics and their functionality are discussed. The good functionality of these products recommends them to be used in different food products. However it's also recognized the need of harmonized methods for the determination of those properties.

Instituto Português do Mar e da Atmosfera (IPMA), Av. Brasília, 1449-006 Lisbon, Portugal.
Emails: cpires@ipma.pt; irineu@ipma.pt
* Corresponding author

INTRODUCTION

Protein hydrolysates have gained increasing importance as food ingredients. Their unique properties differ from intact proteins because of enzymatic hydrolysis. As proteins, the functional properties of protein hydrolysates are related to their molecular size, quantum of reacting amino acid residues, distribution of electrostatic charges, hydrophilicity and hydrophobicity, and carbohydrate and mineral interaction. However, they are also affected by the physicochemical properties of the parent protein used as substrate, the specificity of the protease, the hydrolysis conditions (temperature, pressure, pH, ionic strength, water activity, and solvent polarity), the reaction time, and the combined treatment (enzymatic hydrolysis combined with another enzymic or nonenzymic modification). In fact, as referred by Mahmoud (1994), the peptide hydrolysis lead to an increase in hydrophilicity and net charge, a decrease in molecular size and a change in the molecular structure of the parent protein. However, the functional properties of protein hydrolysates is much more complex as their ability to interact with other constituents in a food formulation depends on the composition and processing parameters employed in food manufacturing.

Several reviews on the functional properties of protein hydrolysates have been published (Schwenke 1997, Haard 2001, Darewicz et al. 2006, Foeding and Davis 2011) but for hydrolysates obtained from fish proteins, the number of reviews is more limited (Kristinsson and Rasco 2000, 2002).

This review gives an overview of the functional properties of fish protein hydrolysates and their importance in food manufacturing. Particular attention has been given to the most recently published papers on this matter.

SOLUBILITY

Solubility is one of the most important functional properties of protein hydrolysates. As mentioned by Mutilangi et al. (1996) enzymatic hydrolysis together with the degradation of proteins into smaller peptides also increases the polar and ionisable groups of protein hydrolysates which in turn promote the interaction with water. Additionally, Gbogouri et al. (2004) also referred that the balance of hydrophilic and hydrophobic forces of peptides is a determinant on their solubility. On the other hand, solubility plays an important role in other functional properties such as foaming and emulsifying because rapid migration and adsorption of the peptides at the interface are critical (Chobert et al. 1988).

Different parts of herring (whole, body, heads, and gonads) were used to prepare FPH using Alcalase. The hydrolysates obtained from the whole body, and heads presented solubility above 70% and that from gonad herring hydrolysate was about 56% (Sathivel et al. 2003). Gbogouri et al. (2004) optimized the preparation of salmon by-products hydrolysates using response surface methodology. The solubility of salmon hydrolysates was higher than 75% over a pH range between 3 and 11, being higher than 90% at pH 6 to 7 and low at pH 3 to 4. A positive relationship between solubility and DH value was also observed. Sardinella by-products were also used to prepare protein hydrolysates which presented high solubility (Souissi et al. 2007). It was also shown an increase of solubility with increasing DH. Round scad protein hydrolysates prepared with Flavourzyme had excellent solubility (99%) which suggested being due to the size reduction and the formation of smaller, more hydrophilic and more solvated polypeptide units (Thiansilakul et al. 2007). Yellow stripe trevally hydrolysates prepared with Alcalase (HA) and Flavourzyme (HF) were soluble over a wide pH range (2–12) with more than 85% solubility and the lowest solubility was recorded at pH 4.0 (Klompong et al. 2007). Both hydrolysates (HA and HF) with higher DH were more soluble than those with lower DH. Pacific whiting hydrolysates obtained with Alcalase had high solubility in distilled water (over 97%) (Pacheco-Aguilar et al. 2008). Dong et al. (2008) reported the preparation of protein hydrolysates from silver carp using Alcalase and Flavourzyme. They also presented high solubility (above 75%) at pH between 2 and 11. These authors also observed an increase in peptide solubility with lower molecular mass. Giménez et al. (2009) used sole and squid skin to prepare protein hydrolysates using Alcalase. Both hydrolysates were highly soluble at pH 7.0 (over 95%) due to the generation of low molecular peptides by hydrolysis. In the preparation of protein hydrolysates from catfish skin with Alcalase the soluble and insoluble fraction were recovered (Yin et al. 2010). The soluble fraction had much higher nitrogen solubility (99.8%) than the insoluble fraction (2.4%). The high nitrogen solubility of the former fraction indicated potential applications in formulated food systems. Foh et al. (2010) prepared tilapia protein hydrolysates with Alcalase, Flavourzyme and Neutrase. The solubility profile of hydrolysates as a function of pH exhibited a U-shaped curve. The Alcalase hydrolysates had the highest solubility values (above 95%) at alkaline pH.

In a second work Foh et al. (2011) also reported the preparation of tilapia hydrolysates using Alcalase. The solubility of hydrolysates versus pH also showed a U-shaped curve having higher solubility values at both alkaline and acidic pH levels. Nalinanon et al. (2011) obtained protein hydrolysates from the muscle of ornate threadfin bream using skipjack tuna pepsin soluble. All hydrolysates exhibited solubility higher than 70% over a wide pH range (3–9). Dos Santos et al. (2011) reported the preparation of hydrolysates from the muscle of the undervalued fish bluewing seaborin using Alcalase and

Flavourzyme. The maximum solubility at pH 9 was 42% for hydrolysates obtained with Alcalase and 38% for those prepared with Flavourzyme. Naqash and Nazeer (2011) evaluated the effect of three proteases (papain, pepsin and trypsin) on the solubility of protein hydrolysates obtained from pink perch. The results obtained evidenced the effect of protease type on the solubility of the hydrolysates. Thus, those produced by trypsin had the highest solubility (98%) compared to pepsin (77%) and papain (74%) hydrolysates. Sampath Kumar et al. (2011) reported the preparation of protein hydrolysates from different body parts (skin, muscle, bone and viscera) of horse mackerel and croaker. The hydrolysis was done through an *in vitro* gastrointestinal digestion procedure using commercial pepsin, trypsin and α-chymotrypsin. All the hydrolysates obtained were soluble over a pH range between 2 and 10 with more than 65% solubility. The lowest solubility of all hydrolysates was recorded at pH 4. Khaled et al. (2011) reported the preparation of sardinelle muscle hydrolysates using three different crude enzymes preparations. The sardinelle hydrolysates were more than 65% soluble at pH ranging from 2.0 to 12.0. The solubility of all hydrolysates was minimum at pH 5 and the hydrolysate obtained with protease from *Bacillus pumilus* showed the highest DH and was also more soluble. Muzaifa et al. (2012) also observed that hydrolysates prepared from fish by-products using Alcalase were more soluble (*ca.* 81%) than those obtained with Flavourzyme (approximately 79%). The autolysate prepared from Pacific whiting had the lowest solubility (82.5%) at pH near the pI of muscle proteins (around pH 4.0–5.0). The solubility was improved to 100% for pH values of 7.0 and 10.0 (Mazorra-Manzano et al. 2012). According to the authors the low solubility at pH 4.0 could be due to the low DH (4.43%) of this autolysate. Amiza et al. (2012) used cobia frames to prepare protein hydrolysates using Alcalase. The solubility of the hydrolysates were in the range of 85–86% and it was not affected by DH.

The preparation of FPH from protein isolates has been followed in several works as an alternative to the direct hydrolysis of the raw material. Choi et al. (2009) reported the preparation of croaker protein hydrolysates with Protamex and Flavourzyme using protein isolates obtained by the pH shift method. Two hydrolysate fractions, a soluble and an insoluble, with different DH were separated. The solubility of the soluble fraction was higher than that of the insoluble fraction at 0% to 3% NaCl and at pH in the range between 2 and 10. Wachirattanapongmetee et al. (2009) also prepared catfish frame hydrolysates from protein isolates obtained by the same method as well as the conventional method. Water and NaCl (0.6 M) solubility of both hydrolysates increased with increasing DH and were markedly superior to those of protein isolates and soy protein isolate. Geirsdottir et al. (2011), also using blue whiting proteins recovered by the pH shift method, prepared protein hydrolysates with Alcalase. They observed a solubility increase with increased DH which was expected as higher DH decreases peptide size and exposes the hydrophilic groups of the amino acids to the solvent. Cape hake by-products hydrolysates were prepared using three different methodologies (direct hydrolysis of fish by-products, hydrolysis of alkaline solubilised proteins and hydrolysis of proteins recovered by the pH shift method) (Pires et al. 2012). All hydrolysates presented a high solubility ranging from 68% to 76% and no significant differences within them were observed.

Balaswamy et al. (2011) prepared protein hydrolysates from catla roe using Alcalase. Before hydrolysis, the roes were previously deffated with isopropanol alcohol and dried (RPC1) or dried and deffated with the same alcohol (RPC2). The protein solubility of the roe hydrolysates was 70.5–99% over a wide pH range (2-12). It was also showed that the solubility of protein hydrolysates obtained from RPC2 was higher than that of hydrolysates prepared from RPC1. The results indicated that the method of roe processing was critical to obtain highly soluble catla roe protein hydrolysates. The protein hydrolysates obtained from meriga eggs showed protein solubility above 72.4% over a wide pH range (Chalamaiah et al. 2010). Balti et al. (2010) showed that the solubility of protein hydrolysates prepared from cuttlefish by-products increased with increasing DH over a pH range between 3 and 10.

EMULSIFICATION/EMULSIFYING PROPERTIES

The emulsifying properties of FPH are related to their surface properties. A pre-requisite for adsorption of a protein on the oil-water interface is the presence of hydrophobic patches on the surface of the protein molecule (Kristinsson and Rasco 2000). Hydrolysates are surface-active compounds and promote oil-in-water emulsion because of their hydrophilic and hydrophobic groups with their associated charges. The emulsifying properties of hydrolysed proteins are influenced by the extent of hydrolysis. The emulsifying properties generally decreased as the percentage of degree of hydrolysis increased due to the formation of small peptides with less surface activity and less hydrophobicity (Kristinsson and Rasco 2002). To possess good emulsifying and interfacial properties it has been suggested that peptides should not be smaller than 20 residues (Lee et al. 1987). A relationship has been found between solubility and emulsifying properties, as hydrolysates with high solubility can rapidly diffuse and adsorb at the interface (Klompong et al. 2007). The dependence of emulsifying activity on the concentration of protein by adsorption kinetics was also previously reported by Kinsella (1976). Emulsifying capacity (EC) and emulsifying stability index (ESI) are generally used to measure the ability of protein hydrolysates to form and stabilise emulsions. Emulsifying capacity is defined as the volume of oil that can be emulsified by the protein hydrolysate before the collapse of emulsion occurs (Kinsella 1976). However, EC is not solely a property of the protein under test but rather is a property of the emulsion system, the equipment and method used to produce emulsions. Thus, Pearce and Kinsella (1978) proposed a method for measuring emulsifying properties of proteins based on the turbidity of the emulsion at a wavelength of 500 nm. Emulsifying activity index (EAI) estimates the ability of a protein to aid in the formation and stabilization of a newly-created emulsion by giving units of area of the interface that is stabilized per unit weight of protein.

Emulsifying capacity reported for whole herring and herring by-product hydrolysates obtained using Alcalase was in the range of 55.5 to 100.3 mL of oil/g of protein (Sathivel et al. 2003). The EC of salmon by-products hydrolysates prepared by Gbogouri et al. (2004) increased with hydrolysate concentration (0.02% to 0.20%) to a maximum and then decreased. These authors showed that high DH does not improve EC. Šližytė et al. (2005) reported the preparation of hydrolysates from cod by-products with Flavourzyme and Neutrase. According to these authors the most significant factor influencing the emulsifying capacity of these protein hydrolysates was the amount of added water before hydrolysis. The EC of sardinella protein hydrolysates obtained by Souissi et al. (2007) decreased with the increase of DH and the hydrolysate with 6.62% DH had a higher EC than casein and the undigested fish proteins. EAI of round scad protein hydrolysates prepared with Flavourzyme decreased with increasing hydrolysate concentration from 0.1% to 3.0%. However, no differences were observed in ESI when the hydrolysate concentrations increased up to 0.5% (Thiansilakul et al. 2007). Wasswa et al. (2007) also observed that EC of grass carp skin hydrolysates decreased with increasing protein hydrolysis. As highlighted by the authors, smaller peptides are less effective in stabilizing the emulsions. Klompong et al. (2007) reported that the EAI and ESI of protein hydrolysates from yellow stripe trevally meat using Alcalase and Flavourzyme decreased with increasing DH. It was suggested that the peptides with low MW may not be amphiphilic enough to exhibit good emulsifying properties. Pacheco-Aguilar et al. (2008) studied the effect of DH and pH on the EC of hydrolysates produced from Pacific whiting using Alcalase. These authors showed that pH was the major factor to be considered in the evaluation of EAI. Emulsion stability index of these hydrolysates showed a tendency to increase at higher pHs, but this property was not affected by the DH. As reported by Choi et al. (2009) the insoluble hydrolysate fraction of croaker protein hydrolysates had higher EAI than the soluble fraction which was comparable to egg white and bovine plasma protein, while that of soluble fraction was comparable to these two proteins. The authors also concluded that EAI of hydrolysates depends on protease type and DH. The EAI of catfish frame hydrolysates obtained by Wachirattanapongmetee et al. (2009) was improved as the hydrolysis time was extended but only for 0.5% enzyme/substrate (E/S) ratio. However, this trend was not observed for higher E/S ratios. The EAI of these hydrolysates was lower than alkali-extracted protein and commercial soy protein isolates but their ESI was generally higher. The EAI of both gelatin hydrolysates obtained from sole and squid skin proteins (Giménez et al. 2009) decreased with increasing hydrolysate concentration. This phenomenon may be explained not only, as mentioned before, by adsorptions kinetics but also to the increase in protein-protein interactions with increasing concentration, resulting in a lower protein concentration at the oil-water interface (Huidobro et al. 1998, Lawal 2004). Tilapia protein hydrolysates prepared with different enzymes presented EC values in the range of 21.40–27.33 mL/0.5g of protein (Foh et al. 2010) where as the hydrolysates prepared with Flavourzyme presented the highest EC. The emulsion stability for the soluble and insoluble fraction of catfish skin protein hydrolysates was 79% and 98%, respectively (Yin et al. 2010). The higher emulsion stability of the insoluble fraction might be due to higher amount of lipid content of this fraction as pointed out by the authors. Both EAI and ESI of hydrolysates from ornate threadfin bream muscle with different DH varied, particularly when the protein concentration increased (Nalinanon et al. 2011). The Alcalase hydrolysates prepared from bluewing seaborin (Dos Santos et al. 2011) showed the highest EAI at pH 3.0, with the most stable emulsion being 0.0275% at pH 11.0. Flavourzyme hydrolysate had the highest EAI at pH 11.0 and the highest emulsion stability of 0.144% at pH 3.0. Naqash and Nazeer (2011) showed that pink perch protein hydrolysates had the lowest EAI and ESI at pH 4.0, which increased as the pH moved away from 4.0. The effect of concentration of various sardinelle protein hydrolysates obtained by Khaled et al. (2011) on the EAI and ESI was evaluated. These indexes of all sardinelle hydrolysates decreased with increasing concentration in the range of 0.1% to 2% which was also found by other authors. The emulsifying properties of a fish protein hydrolysate prepared from Pacific whiting muscle were affected by pH (Mazorra-Manzano et al. 2012). This hydrolysate showed better emulsifying properties at pH 10.0. The results of the emulsifying properties of blue whiting protein hydrolysates prepared by Geirsdottir et al. (2011) were in the same range as the reference samples (soy, whey and milk proteins). They also observed a slight increase of EC with increasing DH which was suggested to result from the increasing hydrolysate solubility which would overcame the eventual EC reduction caused by the reduced peptide sizes. The EC of protein hydrolysates prepared from different body parts of horse mackerel and croaker varied with pH (Sampath Kumar et al. 2011). All hydrolysates exhibited exceptional emulsifying activity and stability. For instance, EAI of viscera hydrolysate from horse mackerel and skin hydrolysate of croaker at pH 10.0 was 250 and 300 m^2/g, respectively. Amiza et al. (2012) reported EC values in the range of 3–12 ml oil/g for cobia protein hydrolysates prepared with Alcalase. EC increased with the decrease of the extent of hydrolysis. Muzaifa et al. (2012) observed a decrease of the EAI and ESI with increasing concentration of FPH prepared from fish by-products using Alcalase and Flavourzyme. However, the FPH produced with Alcalase had higher EAI and ESI compared to FPH produced with Flavourzyme. Pires et al. (2012) reported EAI values for FPH prepared from hake by-products by different methodologies using Alcalase. The EAI values were in the range 5.6–7.7 m^2/g and did not seem to correlate with their corresponding DH values.

Enzymatic hydrolysis of cuttlefish by-products (Balti et al. 2010) improved the EAI of hydrolysates and hydrolysate at DH 5% exhibited lower EAI than the others. As referred by the authors, this improvement in emulsifying properties could be explained by the liberation of medium-molecular weight peptides from the native protein which would enhance the flexibility of the peptides, particularly in the hydrolysate with 10% DH. The emulsifying capacity of protein hydrolysates prepared from meriga eggs by Alcalase and papain were 4.25 and 5.98 ml oil/g hydrolysate, respectively (Chalamaiah et al. 2010). The lower EC of Alcalase hydrolysate was explained by the authors as a result of the higher DH of this hydrolysate when compared to papain hydrolysate.

FAT ABSORPTION/FAT BINDING CAPACITY

This property that expresses the quantity of oil directly bound by the protein is an important functional characteristic of ingredients used in the meat and confectionery industries (Gbogouri et al. 2004). Bulk density of the protein (Kinsella 1976), degree of hydrolysis (Kristinsson and Rasco 2000) and the specificity of the hydrolysing enzyme (Haque 1993) affect the ability of hydrolysates to bind fat.

The protein hydrolysates prepared from various parts of herring by-products exhibited different fat-binding capacity (Sathivel et al. 2003). Both whole herring hydrolysate and herring head hydrolysates had the highest ability to bind soybean oil (7.3 and 6.1 ml/g protein, respectively). Gbogouri et al. (2004) observed a decrease of fat absorption capacity of salmon by-products hydrolysates with increasing DH. The 11.5% DH salmon showed the best ability to bind corn oil. As mentioned by the authors, the relative drop of fat absorption as the DH increased may result from the physical entrapment of oil. Grass carp skin hydrolysates with DH in the range of 5.02–14.9% showed good oil binding capacity (Wasswa et al. 2007). These authors also observed a decrease in oil binding capacity with DH increase of these hydrolysates. On the other hand, Sousisi et al. (2007) did not obtain any correlation between the fat absorption of sardinella by-products hydrolysates and DH. All sardinella hydrolysates exhibited moderate fat absorption. Limam et al. (2008) studied the functional properties of shrimp head protein hydrolysates which showed good fat absorption capacity (3 mL oil/g protein hydrolysate) which was also higher than that of raw shrimp head protein. The ability of different tilapia protein hydrolysates to absorb oil was evaluated by Foh et al. (2010). The soluble fraction of hydrolysates prepared from catfish skin had a higher fat absorption capacity (9.7 ml/g protein) than the insoluble fraction (3.15 ml/g protein) as reported by Yin et al. (2010). The fat absorption of hydrolysates prepared from fresh mince meat was 2.27, 3.07, and 2.77 ml/g for Alcalase, Flavourzyme and Neutrase, respectively. The hydrolysates prepared from hot water dip mince with the same enzymes showed similar fat absorption results. Dos Santos et al. (2011) prepared hydrolysates from bluewing seaborin also using Alcalase and Flavourzyme. Alcalase hydrolysates presented an oil holding capacity of 4.50 g oil/g protein and Flavourzyme hydrolysates 5.52 g oil/g protein. Amiza et al. (2012) prepared cobia frame hydrolysates with different degree of hydrolysis. The oil absorption capacity of hydrolysates was within the range of 2.4–2.8 ml /g and was not affected by DH of cobia frame hydrolysates.

Choi et al. (2009) showed that the fat absorption capacity of the insoluble fraction (3.0–3.0 g oil/g sample) obtained in the hydrolysis of croaker recovered proteins was higher than that of the soluble fraction (*ca.* 2 g oil/g sample) with high DH. These results suggested that fat absorption capacity depended on DH and the molecular size of the peptide composition of the hydrolysates. The alkali-extracted protein hydrolysate prepared with 3% Protex 6L presented the greatest fat absorption but there is no correlation between fat absorption and DH or hydrophobicity (Wachirattanapongmetee et al. 2009). On the other hand, Geirsdottir et al. (2011) reported that the oil binding capacity of blue whiting protein hydrolysates decreased as DH increased and peptide size diminished All the blue whiting hydrolysates had good oil binding capacity which ranged between 2 and 3.5 g oil/g protein. Fat absorption capacity of hake by-products hydrolysates prepared by different methodologies including proteins recovered by the pH shift method was relatively low and ranged between 0.9 and 1.4 g oil/g hydrolysate (Pires et al. 2012).

Chalamaiah et al. (2010) reported the characterization of protein hydrolysates prepared by hydrolysing fish eggs of meriga using Alcalase and papain as proteolytic enzymes. Papain hydrolysates had higher fat absorption capacity (1.05 g oil/g sample) than Alcalase hydrolysates (0.91 g oil/g sample). These results put into evidence the effect of enzyme used on the fat absorption of protein hydrolysates produced. Fat absorption of hydrolysates prepared from cuttlefish by-products exhibited good fat absorption (Balti et al. 2010). A decrease in fat absorption as the DH increased was observed where the hydrolysates with 5% DH had the highest fat absorption (1.97 ml oil/g hydrolysate), followed by the hydrolysates with 10% DH (1.87 ml oil/ g hydrolysate) and hydrolysates with 15% DH (1.81 ml oil/ g hydrolysate).

FOAMING PROPERTIES

Foam is responsible for the desirable texture of many food products, thus the ability of proteins to form and stabilize foam may be an important criteria for food quality. Fish protein hydrolysates have good foaming properties and make stable foams. They could find applications in fish based soufflés and pâtés for example. The factors that are relevant for foaming are similar to those required for emulsification.

Klompong et al. (2007) reported a decrease of foaming capacity and stability of both Alcalase and Flavourzyme protein hydrolysate prepared from yellow stripe trevally. Foam expansions of round scad protein hydrolysates increased from 23.3 to 70% when hydrolysate concentration increased from 0.1% to 3% (Thiansilakul et al. 2007). The protein hydrolysate with a concentration of 3% showed the highest foam stability within the first 10 min. Souissi et al. (2007) reported that foam capacity of sardinella hydrolysates decreased with increase in pH. Foaming capacity of hydrolysates produced from Pacific whiting was very low with values of 2.8, 2.9 and 3 ml of foam/ ml protein solution at pH 4, 7 and 10, respectively (Pacheco-Aguilar et al. 2008). In contrast, these hydrolysates showed good foam stability except at pH 4.0. Foam ability of squid and sole gelatin hydrolysates increased from 6–7% to 32–34% when hydrolysate concentration increased from 0.5 to 2% (Giménez et al. 2009). In this study only slight differences in foam stability were found at increasing hydrolysates concentration. Protein hydrolysates obtained from pink perch by enzymatic hydrolysis with trypsin, pepsin and papain showed low foaming capacity and stability (Naqash and Nazeer 2011). Foaming properties of these protein hydrolysates were affected by pH and both properties had a minimum value at pH 4.0. Nalinanon et al. (2011) reported foam expansion and foam stability of hydrolysates from ornate threadfin bream muscle with different DH at various concentrations. At the same hydrolysate concentration, foam expansion slightly decreased when DH of hydrolysate was increased. With the same DH, hydrolysate concentration had no impact on foam expansion. Foam stability of ornate threadfin bream hydrolysates also slightly decreased with increasing DH of hydrolysates. Protein hydrolysates prepared from fish by-products had good foaming properties (foam ability and foam stability). Furthermore, these foaming properties increased when hydrolysate concentrations increased from 0.1% to 3%. Pires et al. (2012) showed that foaming capacity of protein hydrolysates prepared from hake by-products were relatively high (75–120%) but the foam obtained presented low stability.

Balti et al. (2010) reported the foaming properties of hydrolysates from cuttlefish by-products by measuring their whippability at various pHs. The results showed an increase in foaming capacity of all hydrolysates obtained as compared with the undigested proteins at all pH values tested but foaming capacity decreased with increasing protein hydrolysis. Foam stability also decreased with increasing protein hydrolysis, except at pH 8.0. Limam et al. (2011) studied the foaming capacity of hydrolysates obtained from shrimp head wastes at different pH. The results showed that this property was not affected by pH.

WATER HOLDING CAPACITY/WATER ABSORPTION CAPACITY

Water holding capacity of proteins is of great importance to the food industry because it affects both economic and sensory attributes of the products. Hydrolysis modifies the ability of proteins to absorb and bind to water molecules. The increase of terminal carboxyl and amino groups produces a substantial effect on the amount of absorbed water and strength of the sorption bond making them hygroscopic. Due to this, FPH have shown to have good ability to bind and hold water, which is a particularly useful property for many formulations (Kristinsson and Rasco 2000).

Zhang et al. (2002) showed that the addition of krill protein hydrolysate to lizard fish myofibrils led to a moderate water and ATPase activity decrease during frozen storage. These results suggested that myofibrillar denaturation may be suppressed by the addition of these protein hydrolysates. Slizyte et al. (2005) evaluated how addition of FPH powders from cod by-products into comminuted fish muscle and freezing the mixture influence the WHC of the system after thawing. The addition of 5% sludge powders resulted in an increase of up to 17% in WHC. A linear relationship between amounts of glycine/arginine, alanine, hydroxyproline and sum of hydrophobic amino acids and WHC was observed in this study.

The water holding capacity of hydrolysates prepared from grass carp skin (Wasswa et al. 2007) increased with increasing DH. These authors concluded that the presence of polar groups such as COOH and NH_2 that increased during enzymatic hydrolysis had a substantial effect on the amount of water absorbed. The water absorption capacity of the insoluble fraction of protein hydrolysates obtained from recovered croaker proteins was considerable higher than the protein hydrolysates of the soluble fraction (Choi et al. 2009). The WHC of cod mince with added protein hydrolysates prepared from frozen cod backbones and fresh increased 12% whereas it increased 16% when the protein hydrolysates added were prepared from fresh cod backbones (Šližytė et al. 2009). The tilapia protein hydrolysates prepared with Flavourzyme presented the highest water absorption capacity (2.77 ml/g) when compared with the hydrolysates obtained from the same fish species but using Alcalase and Neutrase (Foh et al. 2010). Dos Santos et al. (2011) reported an increase in the water absorption capacity of bluewing seaborin protein hydrolysates in relation to raw muscle. This increase was considerably higher (58.22%) in the case of hydrolysates prepared with Flavourzyme. Geirsdottir et al. (2011) showed that the addition of blue whiting protein hydrolysates to cod mince increased its WHC from 82% (control sample) to *ca.* 97%. However, these authors also referred that the DH did not have an effect on the WHC of the hydrolysates. Amiza et al. (2012) reported that the water absorption capacity of cobia frame hydrolysates was in the range of 0.8–1.1 ml/g and DH did not affect this property. The addition of hake FPH to minced gilthead sea bream increased the WHC by about 9% (Pires et al. 2012).

Balti et al. (2010) showed that the water holding capacity of protein hydrolysates prepared from cuttlefish by-products increased with the extent of hydrolysis which could be due the increased presence of polar groups during enzymatic hydrolysis.

SENSORY PROPERTIES

The bitterness of protein hydrolysates is the main sensory disadvantage of these products concerning their utilization as human foods. Nevertheless, Noguchi et al. (1975) isolated several acidic oligopeptides from FPH possessing a flavour-enhancing quality like that of monosodium glutamate. Jao and Ko (2002) also obtained concentrated tuna hydrolysates through ultrafiltration and reverse osmosis showing aroma and umami taste. As reported by Wu and Aluko (2007), the off-flavour bitterness of protein hydrolysates can be attributed mainly to hydrophobic-containing peptides released during protein hydrolysis. According to Ney (1979), the bitterness of a peptide depends on their hydrophobic amino acid residues and molecular weight. The average hydrophobicity (Q value) of a peptide is calculated from the free energy values of the amino acids residues and the number of residues. Thus, a peptide has no bitterness when Q value is below 1300 cal/mol but it will be almost certainly bitter when it exceeds 1400 cal/mol and the molecular weight is less than 6000 Da. If the molecular weight is greater than 6000 Da, the peptide has no bitter flavour even for Q values exceeding 1400 cal/mol. However, some exceptions and limitations in Ney's hypothesis have been pointed out (Ohyama et al. 1988, Lemieux et al. 1992). Nosho et al. (1985) took into consideration the importance of the primary structure of peptides and reported that an increase in bitterness was related to the location of hydrophobic amino acids at the C-terminal and the basic amino acids in the N-terminal. Because the off-flavour bitterness of protein hydrolysates is one of the shortcomings for their wide application in foods, several methods have been developed to remove, eliminate or mask the bitterness of peptides. As reviewed by FitzGerald and O'Cuinn (2006), these methods include absorption of bitter peptides on activated carbon, chromatographic removal using matrices and selective extraction with alcohols. In other procedures bitterness is masked by the addition of polyphosphates, specific amino acids (Asp and Glu), α-cyclodextrins and an admixture of hydrolysates with intact proteins. Enzymatic methods have been also tried to protein hydrolysate debittering. They include the use of exopeptidases (amino-, carboxypeptidases and proline specific exopeptidases), transpeptidation reactions (the plastein reaction) and cross-linking using transglutaminase.

In the review by Kristinsson and Rasco (2000) are presented several references on the relationship between structure, the effect of enzymes on flavour profile and methods of debittering. Liaset et al. (2000) used a two-steps hydrolysis process to minimize the bitterness of FPH prepared from Atlantic cod and salmon frames. In the first-step hydrolysis, the raw material was hydrolysed with pepsin, neutrase and Alcalase and in the second step of hydrolysis Kojizyme was used. The hydrolysate prepared from cod frames with Alcalase (2.5 h) and subsequently by Kojizyme (8.5 h) had the highest nitrogen balance at 10% FPH-N inclusion level in rat diets. On the other hand, Liu et al. (2000) demonstrated that the increase in the bitterness of autolytic extracts of head-viscera mixture of frigate mackerel was related to lipid oxidation. Daukšas et al. (2004) used three different methods (treatment with endopeptidase, extraction with butanol and treatment with cholestyramine resin) to eliminate or reduce bitterness from protein hydrolysates obtained from the digestive and gall bladder of cod with commercial enzymes. The authors concluded that the use of butanol and cholestyramine resin separately or in combination reduced the bitter taste from the protein hydrolysates. They also reported that bitterness was not fully explained by the presence of bile or bitter amino acids/peptides. Nilsang et al. (2005) produced FPH from fish soluble concentrate (FSC) using Flavourzyme and Kojizyme. The hydrolysis conditions were optimized using response surface methodology. The optimum conditions for FSC hydrolysis with Flavourzyme were 50 leucine aminopeptidase Units/g protein, 20% (w/w) substrate concentration, 45°C, and 6 h. Flavourzyme did not enhance the formation of some bitter-taste amino acids such as tryptophan and the hydrolysate obtained with this protease possessed a low bitter taste which was less than that of 1 ppm caffeine solution. Wasswa et al. (2007) reported the utilization of macroporous adsorption resin (MAR) to desalt different fish skin hydrolysates obtained with Alcalase. The use of MAR was effective in reducing the salt content and in the decolorization and fish flavour reduction. Similarly, the bitter taste was also reduced to slightly detectable levels. The effect of freshness on the lipid oxidation and fish odour development in protein hydrolysates obtained from Nile tilapia was studied by Yarnpakdee et al. (2012). The authors concluded that hydrolysates from this species' muscle with reduced fish odour could be prepared by using fresh fish and incorporation of mixed antioxidants during hydrolysis.

REFERENCES

Amiza, M.A., Y.L. Kong and A.L. Faazaz. 2012. Effects of hydrolysis on physicochemical properties of cobia (*Rachycentron canadum*) frame hydrolysates. International Food Research Journal 19(1): 199–206.

Balaswamy, K., P.G. Prabhakara Rao, G. Narsing Rao and T. Jyothirmayi. 2011. Functional properties of roe protein hydrolysates from *Catla catla*. *EJEAFChe* 10(4): 2139–2147.

Balti, R., A. Bougatef, N.E.-H., Ali, D. Zekri, A. Ahmed Barkia and M. Nasri. 2010. Influence of degree of hydrolysis on functional properties and angiotensin I-converting enzyme-inhibitory activity of protein hydrolysates from cuttlefish (*Sepia officinalis*) by-products. J. Sci. Food Agric. 90: 2006–2014.

Chalamaiah, M., G. Narsing Rao, D.G. Rao and T. Jyothirmayi. 2010. Protein hydrolysates from meriga (*Cirrhinus mrigala*) egg and evaluation of their functional properties. Food Chemistry 120: 652–657.

Chobert, J.M., C. Bertrand-Harb and M.G. Nicolus. 1988. Solubility and emulsifying properties of caseins and whey proteins modified enzymatically by trypsin. J. Agric. Food Chem. 36: 674–677.

Choi, Y.J., S. Hur, B.-D. Choi, K. Konno and J.W. Park. 2009. Enzymatic hydrolysis of recovered protein from frozen small croaker and functional properties of its hydrolysates. J. Food Sci. 74 (1): C17–C24.

Darewicz, M., J. Dziuba. and M. Dziuba. 2006. Functional properties and biological activities of bovine casein proteins and peptides. Pol. J. Food Nutr. Sci. 15/56, SI 1: 79–86.

Dong, S., M. Zeng, D. Wang, Z. Liu, Y. Zhao and H. Yang. 2008. Antioxidant and biochemical properties of protein hydrolysates prepared from Silver carp (*Hypophthalmichthys molitrix*). Food Chemistry 107: 1485–1493.

Dos Santos, S.D'A., V.G. Martins, M. Salas-Mellado and C. Prentice. 2011. Evaluation of Functional Properties in Protein Hydrolysates from bluewing searobin (*Prionotus punctatus*) obtained with different microbial enzymes. Food Bioprocess Technol. 4: 1399–1406.

Daukšas, E., R. Šližytė, T. Rustad and I. Storrø. 2004. Bitterness in fish protein hydrolysates and methods for removal. Journal of Aquatic Food Product Technology 13(2): 101–114.

FitzGerald, R.J. and G. O'Cuinn. 2006. Enzymatic debittering of food protein hydrolysates. Biotechnology Advances 24: 234–237.

Foegeding E.A. and J.P. Davis. 2011. Food protein functionality: A comprehensive approach. Food Hydrocolloids 25: 1853–1864

Foh, M.B.K., I. Amadou, B.M. Foh, M.T. Kamara and W. Xia. 2010. Functionality and antioxidant properties of tilapia (*Oreochromis niloticus*) as influenced by the degree of hydrolysis. Int. J. Mol Sci. 11: 1851–1869.

Foh, M.B.K., M.T. Kamara, I. Amadou, B.M. Foh and W. Xia. 2011. Chemical and physicochemical properties of tilapia (*Oreochromis niloticus*) fish protein hydrolysate and concentrate. International Journal of Biological Chemistry 5(1): 21–36.

Gbogouri, G.A., M. Linder, J. Fanni and M. Parmentier. 2004. Influence of hydrolysis degree on the functional properties of salmon byproducts hydrolysates. J. Food Sci. 69(8): C615–C622.

Geirsdottir, M., S. Sigurgisladottir, P.Y. Hamaguchi, G. Thorkelsson, R. Johannsson, H.G. Kristinsson and M.M. Kristjansson. 2011. Enzymatic hydrolysis of blue whiting (*Micromesistius poutassou*); functional and bioactive properties. J. Food Sci. 76(1): C14–C20.

Giménez, B., A. Alemán, P. Montero and M.C. Gómez-Guillén. 2009. Antioxidant and functional properties of gelatin hydrolysates obtained from skin of sole and squid. Food Chemistry 114: 976–983.

Haard, N.F. 2001. Enzymic modification of proteins in food systems. *In*: Z.E. Sikorski [ed.]. Chemical & Functional Properties of Food Proteins. Technomic Publishing Co., Inc., Lancaster, USA pp. 155–190.

Haque, Z.U. 1993. Influence of milk peptides in determining the functionality of milk proteins: A review. J. Dairy Sci. 76: 311–320.

Huidobro, A., P. Montero and A.J. Borderías. 1998. Emulsifying properties of an ultrafiltered protein from minced fish wash water. Food Chemistry 61(3): 339–343.

Jao, C.-L. and W.-C. Ko. 2002. Utilization of cooking juice of young tuna processed into canned tuna as condiments: Effect of enzymatic hydrolysis and membrane treatment. Fisheries Science 68: 1344–1351.

Khaled, H.B., N. Ktari, O. Ghorbel-Bellaaj, M. Jridi, I. Lassoued and M. Nasri. 2011. Composition, functional properties and *in vitro* antioxidant activity of protein hydrolysates prepared from sardinelle (*Sardinella aurita*) muscle. J. Food Sci. Technol. DOI 10.1007/s13197-011-0544-4.

Kinsella, J.E. 1976. Functional properties of proteins in foods: a survey. Crit. Rev. Food Sci. Nutr. 8: 219–280.

Klompong, V., S. Benjakul, D. Kantachote and F. Shahidi. 2007. Antioxidative activity and functional properties of protein hydrolysate of yellow stripe trevally (*Selaroides leptolepis*) as influenced by the degree of hydrolysis and enzyme type. Food Chemistry 102: 1317–1327.

Kristinsson, H.G. and B.A. Rasco. 2000. Fish Protein Hydrolysates: Production, Biochemical, and Functional Properties. Critical Reviews in Food Science and Nutrition 40(1): 43–81.

Kristinsson, H.G. and B.A. Rasco. Fish protein hydrolysates and their potential use in the food industry. pp. 157–181. *In*: N. Fingerman and R. Nagabhushanan [eds.]. 2002. Recent Adv. in Marine Biotechnol. Vol. 7, Sci. Publ. Inc., Enfield, NH.

Lawal, O.S. 2004. Functionality of African locust bean (*Parkia biglobossa*) protein isolate: Effects of pH, ionic strength and various protein concentrations. Food Chemistry 86: 345–355.

Lee, S.W., M. Shimizu, S. Kaminogawa and K. Yamauchi. 1987. Emulsifying properties of peptides obtained from the hydrolysates of β-casein. Agric. Biol. Chem. 51: 161–166.

Lemieux, L. and R.E. Simard. 1992. Bitter flavor in dairy products. II. A review of bitter peptides from caseins: formation, isolation and identification, structure masking and inhibition. Lait (Lyon) 72: 335–382.

Liaset, B., E. Lied and M. Espe. 2000. Enzymatic hydrolysis of by-products from the fish-filleting industry; chemical characterisation and nutritional evaluation. J. Sci. Food Agric. 80: 581–589.

Liam, Z., S. Sadok and A.E. Abed. 2008. Enzymatic hydrolysis of shrimp head waste: functional and biochemical properties. Food Biotechnology 22: 352–362.

Liu, C., K. Morioka, Y. Itoh and A. Obatake. 2000. Contribution of lipid oxidation to bitterness and loss of free amino acids in the autolytic extract from fish wastes: Effective utilization of fish wastes. Fisheries Science 66: 343–348.

Mahmoud, M.I. 1994. Physicochemical and functional properties of protein hydrolysates in nutritional products. Food Technology 58(10): 89–95.

Mazorra-Manzano, M.A., R. Pacheco-Aguilar, J.C. Ramírez-Suárez, G. Garcia-Sanchez and M.E. Lugo-Sánchez. 2012. Endogenous proteases in Pacific whiting (*Merluccius productus*) muscle as a processing aid in functional fish protein hydrolysate production. Food Bioprocess Technol 5:130–137.

Mutilangi, W.A.M., D. Panyam and A. Kilara. 1996. Functional properties of hydrolysates from proteolysis of heat-denatured whey protein isolate. J. Food Sci. 61: 270–303.

Muzaifa, M., N. Safriani and F. Zakaria. 2012. Production of protein hydrolysates from fish by-product prepared by enzymatic hydrolysis. Aquaculture, Aquarium, Conservation & Legislation - International Journal of the Bioflux Society 5(1): 36–39.

Nalinanon, S., S. Benjakul, H. Kishimura and F. Shahidi. 2011. Functionalities and antioxidant properties of protein hydrolysates from the muscle of ornate threadfin bream treated with pepsin from skipjack tuna. Food Chemistry 124: 1354–1362.

Naqash, S.Y. and R.A. Nazeer. 2011. Antioxidant and functional properties of protein hydrolysates from pink perch (*Nemipterus japonicus*) muscle. J Food Sci Technol DOI 10.1007/s13197-011-0416-y.

Ney, K.H. 1979. Bitterness of peptides: amino acid composition and chain length taste of foods. ACS Symp. Ser. 115: 149–173.

Nilsang, S., S. Lertsiri, M. Suphantharika and A. Assavanig. 2005. Optimization of enzymatic hydrolysis of fish soluble concentrate by commercial proteases. Journal of Food Engineering 70: 571–578.

Noguchi, M., S. Arai, M. Yamashita, H. Kata and M. Fujimaki. 1975. Isolation and identification of acidic ologopeptides occurring in a flavor potentiating fraction from a fish protein hydrolysate. J. Agric. Food Chem. 23: 49–53.

Nosho, Y., K. Otagiri, I. Shinoda and H. Okai. 1985. Studies on a model of bitter peptides including arginine, proline and phenylalanine residues. II. Bitterness behavior of a tetrapeptide (Arg-Pro-Phe-Phe) and its derivates. Agric. Biol. Chem. 49: 1829–1837.

Ohyama, S., N. Ishibashi, M. Tamura, H. Nishizaki and H. Okai. 1988. Synthesis of bitter peptides composed of aspartic acid and glutamic acid. Agric. Biol. Chem. 52: 871–872.

Pacheco-Aguilar, R., M.A. Mazorra-Manzano and J.C. Ramírez-Suárez. 2008. Functional properties of fish protein hydrolysates from Pacific whiting (*Merluccius productus*) muscle produced by a commercial protease. Food Chemistry 109: 782–789.

Pearce, K.N. and J.E. Kinsella. 1978. Emulsifying properties of proteins: evaluation of a turbidimetric technique. J. Agric. Food Chem. 26(3): 716–723.

Pires, C.,T. Clemente and I. Batista. 2012. Functional and antioxidative properties of protein hydrolysates from Cape hake by-products prepared by three different methodologies. J. Sci. Food Agric: DOI 10.1002/jsfa.5796.

Sampath Kumar, N.S., R.A. Nazeer and R. Jai Ganesh. 2011. Functional properties of protein hydrolysates from different body parts of horse mackerel (*Magalaspis cordyla*) and croaker (*Otolithes rubber*). Mediterr J. Nutr. Metab, DOI 10.1007/s12349-011-0078-3.

Sathivel, S., P.J. Bechtel, J. Babbit, S. Smiley, C. Crapo, K.D. Reppond and W. Prinyawiwatkul. 2003. Biochemical and functional properties of herring (*Clupea harengus*) byproduct hydrolysates. J. Food Sci. 68(7): 2196–2200.

Schwenke, K.K. 1997. Enzyme and chemical modification of proteins. *In*: S. Damodaran and A. Paraf [eds.]. Food Proteins and their Applications. Marcel Dekker Inc., New York, USA pp. 393–423.

Šližytė, R., E. Dauksas, E. Falch, I. Storro and T. Rustad. 2005. Characteristics of protein fractions generated from hydrolysed cod (*Gadus morhua*) by-products. Process Biochemistry 40: 2012–2033.

Šližytė, R., R. Mozuraityte, O. Martínez-Alvarez, E. Falch, M. Fouchereau-Peron and T. Rustad. 2009. Functional, bioactive and antioxidative properties of hydrolysates obtained from cod (*Gadus morhua*) backbones. Process Biochemistry 44: 668–677.

Souissi, N., A. Bougatef, Y. Triki-Ellouz and M. Nasri. 2007. Biochemical and functional properties of sardinella (*Sardinella aurita*) by-product hydrolysates. Food Technol. Biotechnol. 45 (2): 187–194.

Thiansilakul, Y., S. Benjakul and F. Shahidi. 2007. Compositions, functional properties and antioxidative activity of protein hydrolysates prepared from round scad (*Decapterus maruadsi*). Food Chemistry 103: 1385–1394.

Wachirattanapongmetee, K., S. Tawornchinsombut, T. Pitirit, J. Yongsawatdigul and J.W. Park. 2009. Functional properties of protein hydrolysates prepared from alkali-aided protein extraction of hybrid catfish frame. Trends Research in Science and Technology 1(1): 71–81.

Wasswa, J., J. Tang and X.-H. Gu. 2007. Desalting fish skin protein hydrolysates using macroporous adsorption resin. American Journal of Food Technology 2(5): 406–413.

Wu, J. and R.E. Aluko. 2007. Quantitative structure-activity relationship study of bitter di- and tri-peptides including relationship with angiotensin I-converting enzyme inhibitory activity. J. Pept. Sci. 13: 63–69.

Yarnpakdee, S., S. Benjakul, S. Nalinanon and H.G. Kristinsson. 2012. Lipid oxidation and fishy odour development in protein hydrolysate from Nile tilapia (*Oreochromis niloticus*) muscle as affected by freshness and antioxidants. Food Chemistry 132: 1781–1788.

Yin, H., J. Pu, Y. Wan, B. Xiang, P. J. Bechtel and S. Sathivel. 2010. Rheological and functional properties of catfish skin protein hydrolysates. J. Food Sci. 75(1): E11–E17.

Zhang, N.,Y. Yamashita and Y. Nozaki. 2002. Effect of protein hydrolysate from Antarctic krill on the state of water and denaturation of lizard fish myofibrils during frozen storage. Food Sci. Technol. Res. 8(3): 200–206.

4

Angiotensin I Converting Enzyme Inhibitory Peptides from Fish By-products

Raúl Pérez Gálvez, F. Javier Espejo Carpio, Pedro J. García Moreno, Antonio Guadix* and *Emilia M. Guadix*

ABSTRACT

Fish by-products are an inexpensive and easily available raw material which have been traditionally disposed of or used in animal feed or fertilization, i.e., fish feed, fish oil, organic fertilizers. Driven by the current development in biotechnology, and more specifically in separation and purification processes, many approaches have been proposed to upgrade the protein content of marine sources. Several peptides from fish protein hydrolysates have been found to exhibit numerous biological properties, such as antihypertensive, antithrombotic, immunomodulatory and antioxidative, among others.

The antihypertensive property of fish peptides lies in their ability to inhibit the action of the Angiotensin I Converting Enzyme (ACE), an enzyme naturally present in many biological systems which exerts regulatory control over arterial blood pressure. Some ACE inhibitory

Department of Chemical Engineering, University of Granada, 18071 Campus Fuentenueva, Granada, Spain.
Emails: rperezga@ugr.es; fjespejo@ugr.es; pjgarcia@ugr.es; aguadix@ugr.es; eguadix@ugr.es
* Corresponding author

peptides have been reported to cause a similar effect on blood pressure than that of synthetic drugs, such as captopril, without their adverse side-effects.

This chapter discusses, the current literature on ACE inhibitory peptides from fish sources, with particular focus on wastes and by-products from fish processing. As an example, an *in vitro* ACE inhibitory assay conducted on fish discarded species from Mediterranean fisheries is described.

INTRODUCTION

Hypertension

Hypertension is a disorder characterized by increased blood pressure. According to Kearny et al. (2005), in 2000 26.4% of the adult population had hypertension, which represented 972 million people, 333 million in developed countries and 639 in developing countries. By 2025 the hypertension prevalence of predicted to increase by about 60% until reaching 1.56 billion people. Hypertension is spreading at an alarming rate in developing economies such as India and China.

Hypertension is the main risk factor for cardiovascular diseases (Kannen 2000, Mancia et al. 2007) such as coronary artery disease, arrhythmia, stroke and heart attack. According to World Health Organization (WHO), these diseases are the main causes of death today. Indeed, 17.3 million people died some due to cardiovascular disease or the other in 2008, representing 30% of all yearly deaths. An estimated 7.3 million died due to coronary heart disease and 6.2 million due to stroke. WHO estimates that if this trend remains, by 2030, 23.6 million people will die from cardiovascular diseases. Apart from this, hypertension is also a risk factor for other disorders such as renal diseases (Ruilope 1999).

Among the different drugs which are administered to reduce blood pressure, the main groups are diuretics such as angiotensin converting enzyme inhibitors, calcium antagonists, angiotensin receptor blockers and beta blockers. They can be used alone or in conjunction. The Guidelines for the Management of Arterial Hypertension from European Society of Hypertension (ESH) and the European Society of Cardiology (ESC) provide recommendations about their use.

Angiotensing Converting Enzyme (ACE)

As mentioned earlier, the inhibition of ACE is one of the methods to reduce blood pressure. ACE (EC 3.4.15.1), also known as peptidyl-dipeptidase

A, is a zinc metallopeptidase that is activated by chloride. This enzyme is widely distributed in the human body and can be found in the brain, heart, lungs, liver, kidney, gastrointestinal tract, pancreas, testes, etc. (Murray and FitzGerald 2007). Most of it, however (90%) exists as ectoenzyme in the endothelial cell membrane. Besides the membrane bound form there is a free form of ACE that is present basically in all biological fluids (Sibony et al. 1993).

ACE plays a key role in the regulation of blood pressure (Johnston 1992, Campbell 2003) taking part in important regulatory functions (Fig. 1). In the renin–angiotensin system, ACE releases the dipeptide HL from the C-terminal of a decapeptide without blood pressure activity called Angiotensing I (DRVYIHPFHL). The resulting octapeptide DRVYIHPF, known as Angiotensin II is a strong vasoconstrictor. Moreover, in kinin–kallikrein system ACE degrades bradykinin which loses its vasodilator properties. Inhibition of ACE is therefore a must in the reduction of blood pressure.

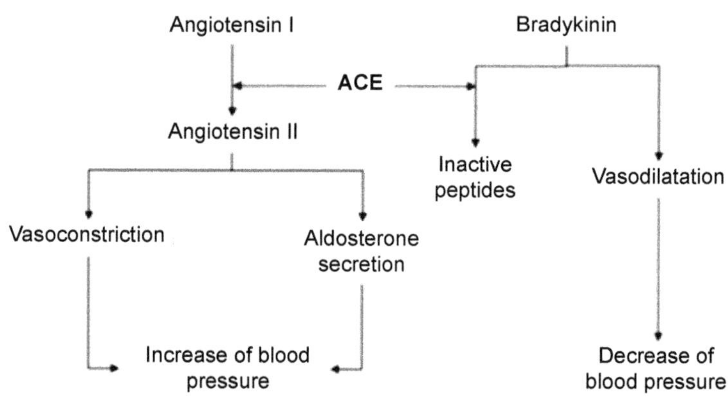

Fig. 1 Effect of Angiotensing converting enzyme (ACE) in blood pressure regulation. Adapted from Li et al. (2004).

The inhibition of ACE

The development of compounds inhibiting ACE began after the discovery those inhibitors by Dr. Ferreira y Rocha e Silva and his team (University of São Paulo) in 1965. They isolated bradykinin-potentiating peptides from *Bothrops jararaca* venom. Following this discovery, numerous compounds capable of inhibiting the enzyme were synthetized. Some of them are now common drugs used to control hypertension, such as captopril, lisinopril, ramipril or enalapril (Pina and Roque 2009). Nowadays, the cost of manufacturing these drugs in developed countries is very high. For instance, in the USA the annual antihypertensive drug costs approximately $ 15 billion (Hong et al. 2008).

However, these drugs have been reported to cause several adverse side effects such as cough, hypotension, increased potassium levels, reduced renal function and skin rashes (Coulter and Edwards 1987, FitzGerald et al. 2004). The patient could even suffer angioedema (swelling caused by leakage from blood vessels) which can be life-threatening (Adam et al. 2002). Therefore, a safer and less expensive alternative to these drugs is desirable. In this regard, antihypertensive peptides derived from natural sources are safer than commercial drugs (Takano 2002, Riordan 2003, Cadée et al. 2007, He et al. 2008, Ishida et al. 2011).

Food Derived Peptides with ACE Inhibitory Activity

Bioavailability of bioactive peptides

Naturally derived peptides must not only be a good ACE inhibitor but they should also reach the cardiovascular system in an active form. Indeed, the potential of ACE inhibition *in vitro* is not always related to its antihypertensive effect *in vivo*. In many cases, peptides with high inhibitory activity *in vitro* have not shown any activity after oral or intravenous administration. For example, YGGY (IC_{50} = 16.2 µM) is a potent ACE inhibitor derived from sake, which has no hypotensive effect after oral administration even if administered in high doses (100 mg/Kg) (Saito et al. 1994). Similar results have been reported by others authors (Fujita et al. 2000) as well.

The capacity of exerting inhibitory activity *in vivo* is related to the resistance of the peptide against the different hydrolysis processes suffered after oral ingestion. Generally, the processes that could alter the bioavailability of the peptides are: gastrointestinal digestion, intestinal absorption and blood transport.

The gastrointestinal process can be simulated *in vitro*, which allows checking the resistance of inhibitory peptides by digestive enzymes (pepsin, trypsin, chymotrypsin and carboxipeptidases). These studies show gastrointestinal digestion as a key process in the determination of ACE inhibitory activity. Active peptides could completely lose their inhibitory capacity after digestion while other inactive peptides could be transformed into potent inhibitors. It has been observed that peptides containing proline are more resistant to the degradation of gastrointestinal enzymes while tripeptides containing proline-proline in the C-terminal are also resistant to proline-specific peptidases (FitzGerald and Meisel 2000).

After gastrointestinal digestion, peptides present in the intestine are commonly introduced into the enterocytes by peptide transporters. Inside the enterocyte, intracellular enzymes hydrolyze the peptides into amino acids which cross the membrane and reach the blood stream (Vermeirssen et al. 2004). The transport of intact peptides (i.e., without undergoing any transformation) through the intestinal wall is not completely elucidated yet. Recent studies have stated that peptide inhibitor transported through the intestine membrane could differ from the traditional peptide transport. Some water soluble peptides, like Valine-Proline-Proline, Isoleucine-Proline-Proline or Valine-Leucine-Proline-Valine-Proline, could cross through intercellular space (Satake et al. 2002, Lei et al. 2008), this transport is known as *paracellular transport*. On the other hand, lipid-soluble peptides seem to be able to cross the membranes by diffusion and the process is called transcellular transport (Knütter et al. 2008).

Finally, peptide stability in blood flow should also be considered when studying bioavailability. Indeed, active peptides should be resistant to enzymes present in the blood stream, such as ACE, which could hydrolyze them, giving rise to new inactive sequences.

Simulated digestion or cell membrane models (Caco-2) could give a good approach to the different mechanisms involved in the bioavailability of inhibitory peptides. This approach has been employed by Bougafet (2008) and Samaranayaka et al. (2010), among others, to predict the *in vivo* antihypertensive activity of fish protein hydrolysates from sardinelle and Pacific hake, respectively.

Taking into account the possibility that inhibitory peptides are degraded by the human body enzymes (gastrointestinal, ACE or enzymes from intestine cells) and based upon Fujita´s classification (Fujita et al. 2000), it is found that there are three kinds of peptides:

A. *Inhibitor type*: these peptides are not modified by the body's enzymes.
B. *Substrate type*: peptides that are hydrolyzed by the body's enzymes yielding peptides with weaker activity.
C. *Pro-drug type inhibitor type*: peptides that become true inhibitors after hydrolysis by the body's enzymes.

Groups A and C show antihypertensive activity after oral administration. According to Robert et al. (1999) the longer is the peptide the lower is if potential to reduce blood pressure *in vivo*.

Although simulated digestion and cell membrane models can predict peptide stability after oral administration, they cannot replace *in vivo* models where the ACE inhibitor peptides are directly tested on animal or human subjects. To this end, spontaneously hypertensive rats (SHR) are a proper model to study hypertension and the effect of antihypertensive peptides. Indeed, *in vivo* studies have been conducted on SHR to test the ACE inhibitory activity of peptides from several marine species such as shrimp (Nii et al. 2008), sardine (Otani et al. 2009) or loach (Li et al. 2012), among others.

Human studies represent the best way to determine the antihypertensive properties of ACE inhibitory peptides derived from food. Only a few peptides identified as ACE inhibitors *in vitro* have been studied in humans (FitzGerald et al. 2004, Saito 2008). In many of these human trials, a descent in blood pressure was detected after the ingestion of peptides or hydrolysates with ACE inhibitory activity.

The majority of human studies were conducted with fermented milk, mainly produced with *Lactobacilus helveticus* (Hata et al. 1996, Seppo et al. 2003, Mizushima et al. 2004, Tuomilehto et al. 2004, Jauhiainen et al. 2005). The peptides IPP and VPP present in this fermented milks are potent ACE inhibitory peptides which reduce the blood pressure of hypertensive persons after a treatment of 8–21 weeks. No adverse effects were detected even in high doses (Ishida et al. 2011). Regarding the fish sources, Kawasaki et al. (2002) incorporated the peptide VY, isolated from sardine hydrolysates with alcalase, into a vegetable drink which was tested on normotensive, mild hypertensive and hypertensive subjects, evaluating both the decrease in systolic and diastolic blood pressure, as well as the presence of side-effects.

Determination of ACE-Inhibitory Activity

There are a several methods to measure the inhibitory activity *in vitro*, these methods permit the evaluation of the potential antihypertensive effect of the compounds assayed. This information is rather useful in order to select between several hydrolysates or peptides in the preliminary stages of investigation. However, *in vivo* studies should be performed to confirm the antihypertensive activity.

In vitro assays

These tests are generally based on the measurement of the activity of ACE when hydrolysing a synthetic substrate. The substrates are usually di- or tri-peptides modified in order to allow the determination of enzyme activity by spectrophotometric, fluorometric, radioisotopic and chromatographic methods (Vermeirsen et al. 2004). Spectrophotometric methods are the most widely employed, since they are simpler and present less disadvantages than fluorometic methods (Holmquist et al. 1979). Among them, the method originally proposed by Cushman and Cheung and the method developed by Holmquist are the most important.

The assay proposed by Cusman and Cheung (1971) is based on the hydrolysis by ACE of hippuryl histidyl leucine (HHL) yielding hipuric acid (HA) and the dipeptide histidyl leucine. The optimum conditions of the assay were determined for pH 8.1–8.3 at a chloride ion concentration of 300 mM and 5–10 mM of substrate (HHL). The hydrolysis of the substrate by ACE is developed at 37ºC, after 30 min the reaction is stopped by adding 1N HCl. The hipuric acid generated is extracted by ethyl acetate, later, the solvent is evaporated and HA is suspended in water and quantified by measuring the absorbance at 228 nm. The amount of hipuric acid generated is related to the ACE activity. The production of hipuric acid from HHL is lineal at least during the first 30 minutes when the concentration of enzyme is higher than 12 mU (Cushman and Cheung 1971). This method has some disadvantages, since the number of stages involved is high and the completion of the whole analysis requires a substantial amount of time. However, the extraction stage is the worst drawback, since the solvent extracts not only the product of reaction (HA) but also it could dissolve the substrate (HHL). Due to the strong absorbance at 228 nm of HHL (Cushman and Cheung 1971, Wu et al. 2002), its extraction would suppose an important source of error.

On the other hand, the assay proposed by Holmquist et al. (1979) was adapted in order to measure the ACE-inhibitory capacity of peptides (Vermeirssen et al. 2002). The assay used N-[3-(2-Furyl)acryloyl]-L -phenylalanyl-glycyl-glycine (FAPGG) as substrate, which is hydrolyzed by ACE yielding N-[3-(2-Furyl)acryloyl]-L–phenylalanyl (FAP) and the dipeptide glycyl-glycine (GG). This reaction produces a drop in absorbance which can be related to the ACE activity. The experimental conditions reported in literature are different, but generally pH is set between 8.1 and 8.3, and the concentration of NaCl is adjusted at 0.3M to enhance the enzyme activity (Cushman and Cheung 1971, Holmquist et al. 1979). The absorbance is monitored at 340 nm during 30 minutes. It is important to employ a substrate concentration above 0.3M in order to have a linear decrease in the absorbance (Holmquist et al. 1979). A recent modification permits to implement this method in micro plate reader (Shalaby et al. 2006). This new approach reduces the volume of reagents and therefore the cost (reagents plus enzyme), and also permits to analyze several samples at the same time.

In order to compare the inhibitory activity exhibited by different peptides, the IC_{50} values are commonly employed. This parameter is defined as the concentration of inhibitor which reduces ACE activity down to 50%. Low values of IC_{50} means good inhibitory activity since a low concentration of the compounds is enough to reduce at 50% the activity of the enzyme.

In vivo assays

In many cases, a peptide with a good ACE inhibitory activity *in vitro* may not have any effect reducing blood pressure (FitzGerald et al. 2004). As mentioned above, *in vitro* assays do not take into account the transformations which could suffer the peptides within the organism. In order to have a more precise knowledge about the antihypertensive activity of an ACE inhibitory peptide, it is tested in spontaneously hypertension rats (SHR) which constitute an accepted model for human hypertension. After that, human studies would be the last step to check the safety and effectiveness of peptides.

Temperature and humidity are normally controlled in rats' studies, in order to maintain optimal conditions for the study. The peptide is administered dissolved in a matrix such as saline solution or fermented milk. Three kinds of rats are usually employed:

(a) *Positive control*: rats fed with a well-known antihypertensive drug like captopril.
(b) *Negative control*: rats fed with the matrix without any extra (antihypertensive) component.
(c) *Study group*: rats fed with the peptide or hydrolysate in study.

In short-term studies, rats are fed with the established dose of the peptide, the blood pressure is measured before and within a period of 8 hours after the intake. In contrast, in long-term experiments, the dosage is orally administrated during several days and blood pressure is monitored before and throughout the experiment. The most common method used to measure blood pressure is a non-invasive blood pressure methodology which employs a tail-cuff placed on the tail to occlude the blood flow. In some cases, the animals are sacrificed once the treatment is finished in order to measure levels of antihypertensive peptides and the activity of angiotensin converting enzyme in different parts of the body: aorta, lung, kidney, heart and brain (Masuda et al. 1996, Sánchez et al. 2011). All these measurements give a clue to the effectiveness of food peptides or hydrolysates in the treatment of hypertension.

The last stage to claim the inhibitory effect of a known peptide is to check their effect on human beings. The subjects in these trials are people with hypertension disorders who participate voluntarily. People with health problems like disorders in hepatic function or problems in renal, cardiovascular or pulmonary systems are not allowed to take part in the trials (Hirota et al. 2007). Subjects without hypertension could also be included in the studies in order to check the consequences of inhibitory peptides intake in people with normal pressure.

Human trials use double-blind experiments, which avoid both experimenter bias and placebo effects. The subjects are divided randomly into two groups; one receives the ACE inhibitory compound and the other group a placebo. Both the active samples and the placebo are usually provided as pills and the participants should take them according to the indications of the researchers.

The effect on blood pressure was monitored for several weeks. The measurements of blood pressure should be made in the same conditions by following a procedure previously established in order to have comparable results. The weight of the subject is also monitored throughout the study and other analyses such as electrocardiograms or blood and urine analysis are commonly done at the beginning and end of the experiment.

REVIEW OF ACE INHIBITORY PEPTIDES ORIGINATING FROM FISH

Fish by-products as Sources of Bioactive Peptides

Fish processing generates variable amounts of by-products which are susceptible for valorisation in order to isolate peptides exhibiting ACE inhibition activities. These by-products are present in large quantities and are a low-cost source.

The fish industry involves a varied number of processing operations to transform the raw material in edible products directly intended for human or animal consumption. Fishery waste on land is increasing nowadays driven by two factors. First, greater processing of fishery products, which leads greater quantities of waste, and second, a greater concentration of waste due to the implantation of new larger industries and fish auctions, rather than smaller ones (Blanco et al. 2007).

Although much of this waste is already being handled, either for fish meal and oil production or treated as urban solid waste, this kind of utilization is inefficient. The present development of technology permits to provide a more profitable use of by-products, and minimize the economic and environmental impact of their disposal. Of the estimated 131 million tonnes of fish produced in 2000 in the world, nearly 74% (97 million tonnes) was used for direct human consumption. The remainder (about 26%) was utilized for various non-food products, mostly for reduction to meal and oil. As a highly perishable material, fish has a significant requirement for processing. In 2000, more than 60% of total world fisheries production underwent some form of processing. An important waste reduction strategy for the industry is the recovery of high valued by-products from fish wastes able to be employed for food, pharmacy or nutraceutical applications.

Commercial processing of marine materials requires the removal of bones, skin, head and viscera (by-products), which account for approximately 60–70% of the fish weight. In general, fish meat and oil in the by-products range widely, but typically represents 20–30% and 5–15%, respectively (Torres et al. 2007). A large amount of the by-products generated in fish processing (up to 60–70% in many cases), are just disposed of or landfilled. These practices should be considered an irresponsible utilization of the natural resources.

A better upgrading of these materials is now possible driven by the implementation of new biotechnological procedures, able to provide a higher added value to the transformed by-products. More specifically, the development of technologies has permitted the extraction, isolation and identification of an increasing number of bioactive compounds from remaining fish muscle proteins, collagen and gelatine, fish oil, fish bone, internal organs and shellfish and crustacean shells. These bioactive compounds can be extracted from the fish matrix and purified with technologies of different degree of complexity (e.g., hydrolysis, solvent extraction or chromatography separations). Such valuable compounds include bioactive peptides, oligosaccharides, fatty acids, enzymes, minerals and other compounds of interest in biotechnological, nutraceutical and pharmaceutical applications.

Fish frames and cutoffs resulting from mechanically deboned fish still present a high content of muscle proteins. They can be recovered from these biomasses by enzymatic hydrolysis, thus obtaining fish protein hydrolysates (FPH). Peptides present in protein hydrolysates from fish muscle protein have exhibited numerous bioactivities such as antihypertensive, antithrombotic, immunomodulatory and antioxidative activities (Gormley 2006, Nagai et al. 2008, Je et al. 2009, among others).

Similarly to other substrates, the antihypertensive activity reported for some fish peptides is due to the inhibition of Angiotensin I Converting Enzyme (ACE). In some cases, this inhibitory effect has been reported to be even stronger than that of many other natural peptides. Moreover, this activity has been tested *in vivo* on spontaneously hypertensive rats (Lee et al. 2010) or even on human subjects (Kawasaki et al. 2002).

ACE Inhibitory Peptides from Fish Protein Hydrolysates

Sardine

The first tests of ACE inhibition activity in fish sources were performed on FPH from sardine (*Sardina pilchardus*) in the mid-1980s (Suetuna and Osajina 1986). The authors reported an IC_{50} value of 3.79 mg/l for sardine hydrolysates with denazyme AP, a protease from *Aspergillus oryzae*. Since then, sardine hydrolysates have been extensively tested for their ACE inhibitory activity, as shown in Table 1. To the authors' knowledge, 14 bioactive peptides have been so far identified, all of them isolated from sardine hydrolysates with alcalase, a protease enzyme from *Bacillus lichenniformis*. The majority of the biopeptides reported in literature were di- and tri peptides, except for the tetrapeptide GWAP, isolated from sardine muscle, which presented an *in vitro* value of IC_{50} of 3.86 µM (Matsufuji et al. 1994). According to the *in vitro* tests, these peptides presented IC_{50} values ranging from 1.63 µM for KW and 330 µM for the tripeptide RFP.

Regarding the antihypertensive activity *in vivo*, only the dipeptive VY was tested on human subjects (Matsufuji et al. 1995, Kawasaki et al. 2000). To this end, a group of 29 volunteers, with mild hypertension, high-normal blood pressure and normal blood pressure were randomly divided into a test and control groups. The members of the test group were administered 195 g of a vegetable drink containing 0.4 mg of VY, once a day within a period of 13 weeks, while the control group was given the same amount of vegetable drink deprived of sardine peptides. The test group presented a significant decrease in SBP, from 142.0 ± 10.3 mmHg at the start of the test to 134.4 ± 11.1 mmHg during the first week of the test period. This value remained until the end of the experiment (13 weeks).

Bordenave et al. (2002) tested the *in vitro* ACE inhibition of autolysates produced from fish canning wastes (sardine, cod and shrimp). The authors reported that 50 µg of sardine autolysate powder presented 30% inhibition of the ACE activity, compared to the 57% inhibition observed for shrimp autolysate.

Bonito

Karaki et al. (1993) isolated the active peptides IRPVQ (IC_{50} =1.4 µM) and GVYPHK (IC_{50} =1.6 µM) from an autolysate from bonito liver and intestine. Both peptides were tested on SHR, which were orally administered a dose of 100–200 mg/kg body weight, showing an effective reduction in the blood pressure.

Fujita and Yoshikawa (1999) isolated eight ACE inhibitory peptides from a thermolysin-digest of Katsuo-bushi, a traditional Japanese food prepared from dried bonito fillets. The authors concluded that one these peptides, LKPNM (IC_{50}=2.4 µM), is a prodrug-type ACE inhibitory peptide which leads to the tripeptide LKP (IC_{50}=0.32 µM) after hydrolysis with ACE.

With regard to the *in vivo* assays, both peptides were tested on SHR, reporting average values of reduction in the maximum systolic blood pressure (SBP) of 23 and 18 mmHg after oral administration of 10 mg/kg of LKPNM and LKP, respectively. Interestingly, both peptides presented an *in vivo* antihypertensive activity much higher than that determined from the

Table 1 Review of ACE inhibitory peptides from fish muscle and by-products.

Species	Parent protein	Amino acid sequence	Enzyme	IC_{50} (µM)	SBP[a] (SHR) mm Hg	SBP[b] (Human) mm Hg	Reference
Sardine	head, viscera		autolysate	30% ACEI			Bordenave et al. (2002)
	muscle	KY	alcalase	1.63			Matsufuji et al. (1994)
	muscle	VY	alcalase	10			Matsufuji et al. (1994)
	muscle	IY	alcalase	10.5			Matsufuji et al. (1994)
	muscle	LY	alcalase	38.5			Matsufuji et al. (1994)
	muscle	VY	alcalase	43.7	7	8	Matsufuji et al. (1995) and Kawasaki et al. (2000)
	muscle	MF	alcalase	44.7			Matsufuji et al. (1994)
	muscle	RY	alcalase	51			Matsufuji et al. (1994)
	muscle	YL	alcalase	82			Matsufuji et al. (1994)
	muscle	MY	alcalase	193			Matsufuji et al. (1994)
	muscle	AKK	alcalase	3.13			Matsufuji et al. (1994)
	muscle	GRP	alcalase	20			Matsufuji et al. (1994)
	muscle	RVY	alcalase	205			Matsufuji et al. (1994)
	muscle	RFP	alcalase	330			Matsufuji et al. (1994)
	muscle	GWAP	alcalase	3.86			Matsufuji et al. (1994)
Bonito	actin	IY	thermolysin	2.3	19		Yokoyama et al. (1992) and Fujita and Yoshikawa (1999)
	muscle	LKP	thermolysin	0.32	18	12.55	Fujita and Yoshikawa (1999)
	muscle	IKP	thermolysin	6.9	20		Fujita and Yoshikawa (1999)
	actin	FQP	thermolysin	12			Yokoyama et al. (1992)

Table 1 contd....

Table 1 contd.

Species	Parent protein	Amino acid sequence	Enzyme	IC_{50} (μM)	SBP[a] (SHR) mmHg	SBP[b] (Human) mmHg	Reference
	muscle	LKPNM	thermolysin	2.4	23	12.55	Yokoyama et al. (1992) and Fujita and Yoshikawa (1999)
	actin	IWHHT	thermolysin	3.5	26		Fujita and Yoshikawa (1999)
	actin	IWHHT	thermolysin	5.8			Yokoyama et al. (1992)
	actin	ALPHA	thermolysin	10			Yokoyama et al. (1992)
	muscle	IKPLNY	thermolysin	43			Yokoyama et al. (1992)
	muscle	KYGLYP	thermolysin	62			Yokoyama et al. (1992)
	actin	IVGRPR	thermolysin	300	25		Iroyukifujita et al. (2000)
	actin	IVGRPRHQG	thermolysin	2.4			Yokoyama et al. (1992)
Tuna	muscle	PTHIKWGD	acid				Kohama et al. (1988)
	muscle	WPEAAELM-MEVDP	pepsin	21.6	18		Qian et al. (2007)
	frame	GDLGKTTTV-SNWSPPKWK-DTP	pepsin	11.3	21		Lee et al. (2010)
Salmon	muscle	VW	thermolysin	2.5			Ono et al. (2003)
	muscle	IW	thermolysin	4.7			Ono et al. (2003)
	muscle	MW	thermolysin	9.9			Ono et al. (2003)
	muscle	LW	thermolysin	17.4			Ono et al. (2003)
	muscle	WM	thermolysin	96.6			Ono et al. (2003)
	muscle	WA	thermolysin	277			Ono et al. (2003)

Alaska pollack	skin	GLP	a+p+c[c]	2.65			Byun and Kim (2001)
	skin	GPM	a+p+c[c]	17.1			Byun and Kim (2001)
	frame	FGASTRGA	pepsin	14.7			Je et al. (2004)
Shark	muscle	CF	SM98011	1.45			Wu et al. (2008)
	muscle	EY	SM98011	1.98			Wu et al. (2008)
	muscle	FE	SM98011	2.68			Wu et al. (2008)
Sea bream	scales	VY	alkaline protease	16	≈ 20		Fahima et al. (2004)
	scales	GY	alkaline protease	265			Fahima et al. (2004)
	scales	GF	alkaline protease	708			Fahima et al. (2004)
	scales	VIV	alkaline protease	7.5			Fahima et al. (2004)
Cod	heads		alcalase	30% ACEI			Bordenave et al. (2002)
Sardinelle	heads, viscera		sardine viscera extract	0.81[d]			Bougatef et al. (2008)
Yellowfinsole	frame	MIFPGAGGP-EL	chymotrypsin	28.7	22		Jung et al. (2006)
Jellyfish	collagen	DPGLEGAHG	papain	3.8[d]			Joon et al. (2011)

[a]Maximum decrease in systolic blood pressure observed in spontaneously hypertensive rats.
[b]Maximum decrease in systolic blood pressure observed in human subjects.
[c]Sequential enzymatic treatment: alcalase + pronase E + collagenase.
[d]Value of IC_{50} in mg/mL, determined for the crude hydrolysate without any purification step.

in vitro assays. Moreover, the whole hydrolysate was orally administered to a small group of 30 hypertensive and borderline hypertensive human subjects, reporting an average SBP reduction of 12.55 mm Hg. Consequently, this hydrolysate has been approved as Foods for Specified Heath Use (FOSHU) by the Japanese health authorities.

Tuna

Qian et al. (2007) isolated an antihypertensive peptide from a pepsin hydrolysate of Bigeye tuna dark muscle. After purification, the structure of the peptide was identified to be WPEAAELMMEVDP, with a molecular weight of 1581 Da and IC_{50} value of 21.6 µM.

The antihypertensive effect of this peptide was tested *in vivo* on SHR. Maximal reduction in SBP was detected at 3 h after oral administration at a dose of 10 mg/kg of animal body weight. The authors suggested that this peptide could be employed as an ingredient for functional food or pharmaceuticals against hypertension and other related diseases.

Lee et al. (2010) prepared hydrolysates from tuna frame protein by employing six proteases: alcalase, neutrase, pepsin, papain, α-chymotrypsin and trypsin. ACE inhibitory activity was determined for all the crude hydrolysates, observing that the peptic hydrolysate exhibited the highest inhibitory potency (28% of ACE inhibition). This hydrolysate was then fractionated into three ranges of molecular weight (below 1, 1–5 and 5–10 kDa) using an ultrafiltration membrane bioreactor system. The fraction between 1 and 5 kDa showed the highest ACE inhibitory activity and was used for subsequent purification by ion exchange FPLC, followed by reverse-phase HPLC. A potent ACE inhibitory peptide (IC_{50}=11.28 µM) was isolated, composed of 21 amino acids (GDLGKTTTVSNWSPPKWKDTP).

Antihypertensive activity of this peptide was evaluated *in vivo* by measuring the decrease of systolic blood pressure (SBP) at 1, 2, 3, 6, and 9 h after oral administration of 10 mg/kg of body weight, employing captopril as a positive control. The maximum descent in SBP, 21 mm Hg was recorded at 6 h after administration. This effect was similar to that obtained with captopril (26 mm Hg) within the same period.

Chum salmon

Ono et al. (2003) studied the ACE inhibitory activity of a hydrolysate of defatted chum salmon muscle with thermolysin. The crude hydrolysate presented an IC_{50} value of 27.9 µg/mL, and was tested for *in vivo* activity in SHR rats. Within 4 h after oral administration of the hydrolysate, at doses of 500 and 2000 mg/kg of animal weight, the SBP was reduced approximately

28 and 38 mmHg, respectively. This effect continued up to 8 h post oral administration.

The active peptides present in the hydrolysate were isolated after two consecutive stages of gel filtration and reverse-phase HPLC chromatography, which permitted to identity six dipeptides (VW, IW, MW, LW, WM and WA) with IC_{50} values ranging from 2.5 to 277.3 µM. VW(2.5 µM) and IW (4.7 µM) were found to be the major peptides responsible for the ACE inhibitory actitivity.

Alaska pollack

Byun and Kim (2001) tested the ACE inhibitory activity of proteolytic digests of gelatin extracts from Alaska Pollack (*Theragra chalcogramma*) skin. To this end, gelatin extracts were successively hydrolyzed by Alcalase, pronase E, and collagenase using a three-step recycling membrane reactor. The filtrate exiting the third membrane reactor (1 kDa MWCO) was composed of peptides ranging from 0.9 to 1.9 kDa and presented an IC_{50} of 0.63 mg/mL. This fraction was further purified by gel filtration, ion-exchange chromatography and reverse-phase HPLC, isolating two novel antihypertensive peptides, GPL (IC_{50} = 2.6 µM) and GPM (IC_{50} = 17.13 µM). The authors suggested that GPL could be employed as antihypertensive agent in the formulation of pharmaceutical preparations.

More recently, Je et al. (2004) isolated the active octapeptide FGASTRGA from a pepsin hydrolysate from Alaska pollack frame protein, which displayed low *in vitro* ACE inhibitory activity (IC_{50}= 14.7 µM).

Shark

Utilization of shark parts other than cartilage or liver is residual, and meat is often discarded in shark processing. Wu et al. (2008) reported ACE inhibitory activity (IC_{50} = 0.4 mg/mL) of the hydrolysate of shark meat with protease SM98011, extracted from the *Bacillus* spp. 98011. Three novel active dipeptides, CF, EY and EF, were isolated and identified from this hydrolysate after successive stages of ultrafiltration (3 kDa MWCO) and reverse-phase chromatography. The values of IC_{50} for the new peptides ranged between 1.45 and 2.68 µm.

Sea bream

Sea bream scales, collected from fish processing factories in Japan, were employed as collagen source to prepare hydrolysates with an alkaline protease (Fahima et al. 2004). Crude hydrolysates presented an IC_{50} value of

0.57 mg/mL, and were purified by successive chromatography separation stages, until isolating four active peptides which were identified as GY, GF, VY and VIV. The administration of the hydrolysate to SHR (300 mg/kg of animal weight) resulted in a significant reduction of the blood pressure after 28 days of treatment, as compared to the control group (174 mm Hg at 28 days).

Miscellaneous fish species

Bordenave et al. (2002) prepared alcalase hydrolysates from cod heads which were tested for their ACE inhibitory activity. The antihypertensive potency of the crude hydrolysates was low, and was increased after purification by reverse-phase HPLC, obtaining a peptidic fraction with 30% of ACE inhibition.

Bougatef et al. (2008) tested the *in vitro* ACE inhibitory activity of several protein hydrolyzates prepared from heads and viscera of sardinelle (*Sardinella aurita*), treated with several proteinases (alcalase, chymotrypsin) as well as two extracts from microbial and visceral origin. All the hydrolysates presented ACE inhibitory activity, with IC_{50} values ranging from 1.24 and 7.4 mg/mL. The hydrolysate prepared with sardine viscera enzymes displayed the maximum ACE inhibition, and was further fractionated by size exclusion chromatography, followed by reverse-phase HPLC until obtaining a purified fraction with IC_{50} of 0.81 mg/mL. A simulated digestion assay was performed on this fraction, showing that the bioactive peptides herein contained were resistant to gastrointestinal proteases.

Lee at al. (2010) investigated the ACE inhibition of the hydrolysates of yellowfin sole (Limanda aspera) frame protein, employing α-chymotrypsin as proteinase. The hydrolysaste was filtered through an ultrafiltration membrane of 5 kDa of molecular weight cut-off (MWCO). From this fraction, a peptide with molecular mass of 1.3 kDa was isolated and sequenced (MIFPGAGGPEL). This active peptide, which presented an *in vitro* IC_{50} value of 22.3 μM, was then tested for its in vivo activity on SHR, reporting a reduction in SBP of 22 mm Hg at 3 hours after administration (10 mg/kg of animal body weight). This reduction was comparable to that attained by the group treated with Captopril, and its effect remained after 9 h.

Giant jellyfish is the target by Korea, Japan and China fisheries, where it is often discarded as by-catch species or for reason of its content of toxins (Lee et al. 2007). Yoon et al. (2011) isolated a novel ACE inhibitory peptides fraction from papain hydrolysate of giant jellyfish collagen, with a molecular weight between 3000 and 5000 Da. After fractionation of the hydrolysate by ultrafiltration (1, 3 and 5 kDa MWCO) and further purification by reverse-phase HPLC, a new ACE inhibitory peptide was identified which presented an in vitro value of IC_{50} of 3.8 μg/mL.

Recent Patents on the Production of ACE Inhibitory FPH

The first approaches to patent a process for obtaining of ACE-inhibitory peptides from fish sources were made by Osajima and his colleagues (Osajima and Ooihi 2005, Osajima et al. 2006). They patented the production of the dipeptide Val-Tyr from sardine as a potent inhibitor. To this end, the raw material was homogenized and treated with an alkaline protease (40–55°C, pH above 8), to obtain a fish hydrolysates which was subsequently purified by means of a peptide-absorbing resine and eluted with a 10% ethanolic solution. This product was employed as an ingredient in food supplements or pharmaceutical preparations. Several *in vivo* tests were performed on both normo and hypertensive subjects, evidencing the hypotensive effect of this concentrate.

Secondly, Ewart et al. (2007) patented a batch process to produce a fish protein hydrolysate presenting ACE-inhibitory activity from salmon leftovers, which comprised at least one peptide from the group consisting of: LAF, LTF, IIF, LAY, IAY, VFY, YAY, VLW, IAW, YAL and YNA. In the process, the salmon frames were de-boned, grinded and homogeneized with demineralized water. The endogenous enzymes were denatured by heating at 70°C for 10 minutes. Protease S Amano enzyme produced by Amano Enzyme USA Company Limited was employed in a ratio of 2.6% w/w. The hydrolysis conditions were set at 50°C and pH 8, and the reaction was allowed to take place during 7 hours resulting in a degree of hydrolysis of about 17%. The aqueous fraction was separated from the oil and solid phases by centrifugation and then it was purified by ethanol precipitation, ultrafiltration or reverse-phase chromatography. Finally, the protein soluble fraction was concentrated with a rotary evaporator and spray-dried to yield a powdered fish protein hydrolysates (FPH). The FPH obtained has been demonstrated to be a potent inhibitor of ACE by determination of the IC_{50} value (58.5 μg/mL), and to reduce mean blood pressure in spontaneous hypertensive rats.

Lately, Hiroyuki (2008) patented a method to produce peptides with an ACE inhibitory activity from different raw material such as mackerel, cod and bonito by-products and employing a different enzyme preparation, thermolysin. The hydrolysis was carried out at 60°C and pH 7 with an E/S ratio of 1%. In this patent, the authors proposed to filter the aqueous fraction through an ultrafiltration membrane of molecular weight cut off 3000 Da. Thus, a concentration lower than 5% of peptides with a molecular weight at least of 5000 Da. IC_{50} values ranging from 30 to 200 μg/mL were obtained for the different fractions produced. The peptides produced having a good colour, no bitterness and a nice after-taste could be used as foods and special foods for health care.

CASE STUDY: ACE INHIBITORY ACTIVITY OF HYDROLYSATES FROM MEDITERRANEAN FISH DISCARDED SPECIES

Background and Objectives of the Research

Spanish fishing vessels in the West Mediterranean Sea (FAO fishing area 37.1.1.) employ trawling and purse seine as main fishing gears, which generate an average discard rate of 23% and 5–10%, respectively (FROM 2008). Discards in West Mediterranean Sea comprise non-target species such as mackerel, which appears as by-catch species in trawling fisheries. Other species, such as sardine and horse mackerel, are mostly discarded due to quota restriction or minimum landing sizes (11 cm for sardine and 12 cm for horse mackerel). Horse mackerel is the commercial species presenting the largest discard rate (5–10%) for the Spanish Mediterranean fisheries employing purse seine (FROM 2008). In the case of sardine, although the average discard rate is estimated at only 0.1–0.2% of the total catches in Mediterranean Spanish fisheries, the total mass of the sardine discarded yearly is significant (Carbonell et al. 1997).

In line with the future regulations of the Europen Commission, banning discards in EU fisheries, an up-grading procedure was assayed on three discarded species: sardine, mackerel and horse mackerel. To this end, these species were employed as raw materials for enzymatic hydrolysis. The final hydrolysates were then tested *in vitro* for their ACE inhibitory activity. This section covers the main results obtained after the enzymatic treatment and the ACE *in vitro* tests.

Materials and Methods

Characterization of the raw material

Sardine (*Sardina pilchadus*), mackerel (*Scomber scombrus*) and horse mackerel (*Trachurus trachurus*) were chosen for this study as species discarded by the Spanish fleet in the West Mediterranean Sea. The raw material was provided by the fishing harbour of Motril (Spain), kept in ice during the transportation and pressed in the same day.

The proximate composition was determined for the three species according to the official methods recognized by the AOAC. Ash and moisture content were determined gravimetrically, while solvent extraction was employed to determine the lipid content of the raw material. Total nitrogen was determined by the Kjeldahl method, and the content of crude protein was then obtained by employing a nitrogen-to-protein conversion factor of 6.25 (AOAC 2006).

Separation of protein fraction

The raw materials were dewatered and defatted prior to the hydrolysis, in order to obtain a protein fraction. To this end, the raw material was preheated at 40°C during 30 minutes by means of a water bath model Digiterm 100 (Selecta, Spain). This step improves the separation of protein and lipid phases. Then, it was fed into an electric press model ESP-K (Sanahuja, Spain) where it was submitted to three consecutive pressing steps until attaining a final pressure of 150 bar. As a consequence of the pressing operation, two main phases were obtained: a press liquor comprising an upper oily layer and an aqueous solution, and a press cake deprived of its lipid and moisture content. The cakes resulting from the pressing operation were then frozen at –20°C prior to their use as substrate for protein hydrolysis.

Enzymes and hydrolysis procedure

The press cakes aimed to obtain peptidic fractions exhibiting ACE-inhibitory activity. To this end, the cakes were hydrolysed employing two serine endoprotease enzymes, one of bacterial origin (subtilisin, EC 3.4.21.62) and other from an animal source (trypsine, EC 3.4.21.4), both donated by Novozymes (Denmark) as Alcalase 2.4L and PTN 6.0S, respectively. Subtilisin hydrolyses peptide bonds in a wide range of specifity, and presents an optimum activity at 50-60°C and alkaline pH (Adler-Nissen 1986). On the contrary, trypsin only cleaves specific peptide bonds at the C-terminal side of lysine and arginine residues, except for Arginine-Proline- and Lysine-Proline-bonds which are normally resistant to proteolysis. It exhibits maximum proteolytic activity at pH between 7 and 9 (Sipos and Merkel 1970) and temperature between 30 and 60°C. According to the stability of both enzymes, the experimental conditions were set at 50°C and pH 8, employing an enzyme-substrate ratio of 3% w/w.

For each experiment, the press cake was grinded and homogeneised with deminarilised water until obtaining a suspension with 3% w/w protein. 200 mL of this suspension were then transferred into a jacketed reactor, where the hydrolysis reaction was carried out. Three enzymatic treatments were assayed: (a) hydrolysis with subtilisin for 2 hours, followed by addition of trypsin until completing 4 h of reaction; (b) tripsine (2 h) followed by subtilisine (2h); (c) 4-hour hydrolysis with simultaneous addition of both enzymes. The degree of hydrolysis, defined as the percentage of available peptidic bonds which are cleaved in the course of the hydrolysis, was calculated according to the pH-stat method, which relates degree of hydrolysis with the base consumption throughout the reaction (Camacho et al. 2001). Indeed, the leave of a peptide bond, under alkaline conditions, release protons to the medium which lower the pH of the reaction mixture. Alkaline conditions can be kept in the course of the reaction by employing an automatic titrator 718 Stat Titrino (Metrohm, Switzerland). According to this method, the degree of hydrolysis (DH) can be related to the amount of base consumed to keep the pH constant during the reaction, as follows:

$$\text{DH (\%)} = \frac{B \cdot N_b}{\alpha \cdot m_P \cdot h_{TOT}} \times 100 \qquad (1)$$

where B (mL) is the volume of base consumed, N_b (eq/L) is the normality of the base, α is the average degree of dissociation of the α-NH_2 amino groups released during the hydrolysis, which was calculated as 0.885 at 50°C and pH 8, m_P (g) is the mass of crude protein in the substrate and h_{TOT} (meq/g) is the number of equivalents of peptide bonds per mass of protein, which was assumed to be 8.6 meq/g for fish protein, as reported in literature (Adler-Nissen 1986).

In order to obtain a large number of hydrolysates presenting different molecular weight distribution, several samples were drawn at different times of reaction for each species and enzymatic treatment. These samples were tested for their ACE-inhitory activity *in vitro*. In total, 150 hydrolysates of the three species considered were analysed. To this end, the samples were heated at 100°C for 15 min to deactivate the enzyme, and were then kept under refrigerated conditions until performing the analysis.

Determination of in vitro ACE-inhibitory activity

The *in vitro* antihypertensive activity of the hydrolysates was determined employing the ACE-inhibitory assay described by Shalaby et al. (2006). The method is based on the hydrolysis of the tripeptide N-[3-(2-Furil) acriloil]-L-fenilalanil-glicilglicina (FAPGG, SigmaAldrich, USA) with the Angiotensin converting enzyme (ACE) from rabbit lung (Sigma Aldrich, USA). The assay was carried out in 96-well microplate at 37°C monitoring the absorbance at 340 nm during 30 minutes by means of a Multiskan FC microplate photometer (Thermo Scientific, Finland). Each sample was analysed in triplicate. To this end, 10 µL of sample were mixed with 10 µL of the ACE solution, of actitivy 0.25 U/mL and 150 µL of a solution 0.88 mM of FAPGG in buffer Tris-HCl 50 mM, pH 7.5 and 0.3 M of NaCl. The decrease of the absorbance was monitored in the course of the reaction, so the ACE-inhibitory activity was calculated by Eq. (2):

$$\text{ACE inhibition (\%)} = \left(1 - \frac{\rho_i}{\rho_0}\right) \times 100 \qquad (2)$$

Where ρ_i is the slope in the presence of inhibitor and ρ_o the slope obtained in the absence of inhibitor (pure water). These slopes were calculated from the values obtained within the interval between 10 to 25 minutes, where a better linearity was achieved.

The highest values of ACE-inbitory activity obtained were then reported as IC_{50}. The IC_{50} value was defined as the concentration of hydrolysate (μg protein/ml) required to produce 50% inhibition of ACE. Each value of ACE-inhibitory activity was reported to the content of soluble protein in the hydrolysates, which was determined using Micro-Lowry assay (Sigma Aldrich, USA).

Results and Discussion

Proximate composition of the species

The proximate composition of the fish discarded species studied is shown in Table 2. The lipid content varied among the species, ranging from 2.5% for mackerel to 7.8% for horse mackerel. The three species presented similar protein content, varying from 16.9% for horse mackerel to 18.4% for mackerel. Although proximate composition of fish varies according to the environmental (fish ground, season), physical (e.g., stage of maturity, sex, spawning cycle) and the nutritional condition of the animal, the protein content does not suffer significant variations (Kocatepe and Turan 2012).

Table 2 Proximate composition in wet basis of sardine, mackerel and horse mackerel.

Species	Moisture,%	Lipid,%	Protein,%	Ash,%
Sardine	71.8	4.4	17.6	3.8
Mackerel	74	2.5	18.4	2.8
Horse mackerel	70.4	7.8	16.9	4.0

Time evolution of the degree of hydrolysis

The evolution of the degree of hydrolysis (DH) in time was studied for the three substrates and the three enzymatic treatments. As an example, Fig. 2 shows the time evolution of the DH of sardine hydrolysates for the three enzymatic treatments assayed.

All the hydrolysis curves were similar to that given as example, presenting a high rate of hydrolysis after the addition of the corresponding enzyme. This rate decreased throughout the reaction until reaching a steady-state phase when no apparent hydrolysis took place. Nevertheless, when trypsin was added, the DH increased to a lesser extent and then rapidly flatted down. This fact was attributed to the narrow specifity of trypsin,

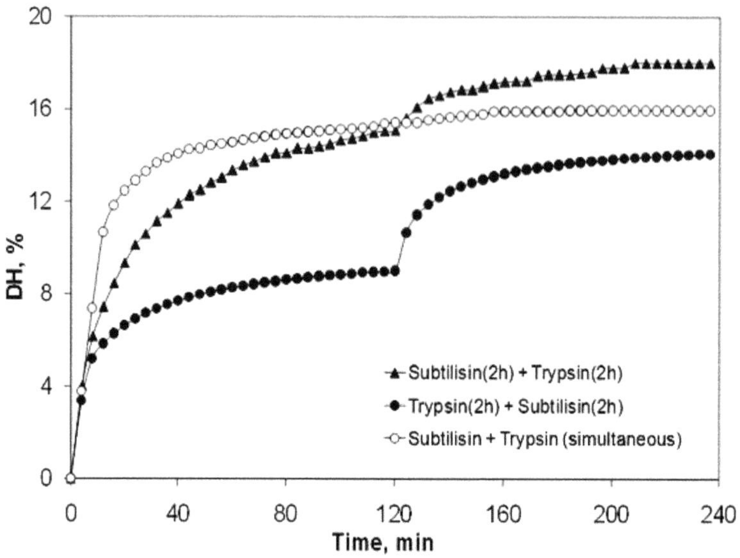

Fig. 2 Time evolution of the degree of hydrolysis for sardine hydrolysates.

which only cleaves peptidic bonds involving arginine or lysine, compared to the broad spectrum of subtilisin (Adler-Nissen 1986).

The DH reached after 4 hours of reaction was different according the species and the treatment. In general, mackerel attained the lowest values of final DH, followed by sardine. Horse mackerel protein presented a best degradability to the enzymes used, reaching the highest levels of DH. Regarding the enzymatic treatments, the sequential reaction subtilisin plus trypsin led to the highest DH for sardine (18%, Fig. 2). In the case of mackerel, the two-step reaction subtilisin plus trypsin and the simultaneous addition of both enzymes led to the maximum DH (14%), while the maximal extent of the hydrolysis (21%) was obtained for horse mackerel after the simultaneous addition of subtilisin and trypsin.

ACE-inhibitory activity

Similarly to the DH, the ACE-inhibitory activity of the hydrolysates was followed at different times of reaction for the three species and the three enzymatic treatments assayed. As an example, Fig. 3 shows the percentage ACE inhibition of sardine hydrolysates against the time for each enzymatic treatment.

It was observed that, for all species and enzymatic treatments, ACE-inhibitory activity increased steeply within the first hour of hydrolysis and then fluctuated around an average value. Furthermore, the addition

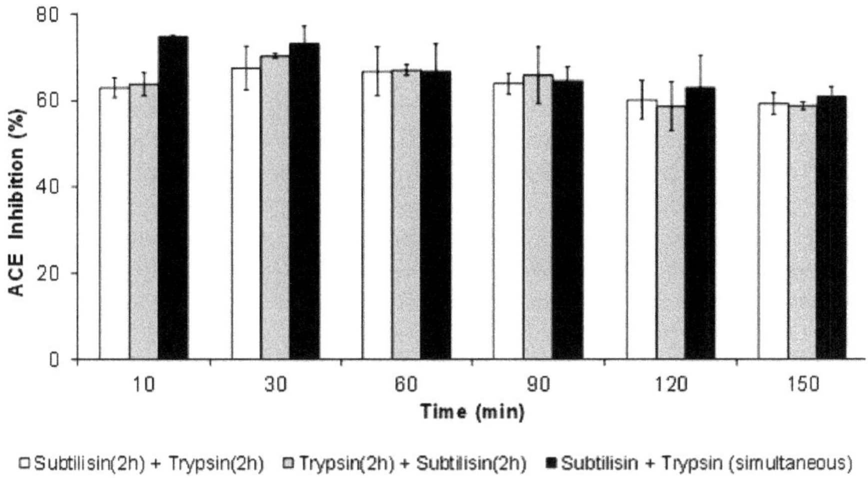

Fig. 3 Time evolution of the ACE inhitory activity for sardine hydrolysates.

of a new enzyme had no significant effect on the ACE inhibition. It was therefore concluded that there is no direct relationship between the degree of hydrolysis and the antihypertensive activity of the hydrolysates. This may be attributed to the fact that the hydrolysates present a wide distribution of peptidic species in constant change due to the proteolitic action. This means that some active peptides are continuously inactivated by the enzymatic action, while new active sequences are generated from previously inactive species.

According to previous literature, ACE inhibition activity is thought to be related with the amino acid sequence, specially the tripeptide sequence at the C-terminal end. Indeed, the presence of hydrophobic amino acid residues in this region seems to favor the ACE inhibitory activity of the peptides, while hydrophilic peptides display weak or no affinity for the ACE active sites (Ryan 2011).

According to Fig. 3, it can be concluded that the three enzymatic treatments assayed did not produce significant differences in the ACE inhibitory activity of sardine hydrolysates. A maximum ACE inhibition value of 70% were obtained, which corresponded to an IC_{50} value of 400 μg protein/mL. This IC_{50} value was higher than the one reported by Matsui et al. (1993) for the same substrate, IC_{50} = 260 μg protein/mL. However, it should be mentioned that no fractionation or purification of the hydrolysate was carried out in this research work, which would considerably improve its final antihypertensive activity.

In the case of mackerel hydrolysates, ACE inhibition values up to 67% were reached after one hour of reaction. The enzymatic treatments assayed showed no significant influence on the antihypertensive activity of the hydrolysates. The sample presenting the highest antihypertensive activity had an IC_{50} value of 345 µg protein/mL, which was the maximum value of ACE-inhibition of the three species studied. Considering the horse mackerel hydrolysates, it was noticed that there is no difference on the ACE-inhibitory activity when employing the sequential reaction subtilisin plus trypsin or the simultaneous treatment with trypsin and subtilisin. ACE inhibition values up to 65% were obtained for both enzymatic patterns, presenting an IC_{50} value of 364 µg protein/mL. However, employing trypsin as the primary enzyme was detrimental to the antihypertensive activity. Indeed, this treatment led to hydrolysates exhibiting lower ACE-inhibitory activities (inferior to 60%) than those obtained after addition of subtilisin. To the authors' best knowledge, no IC_{50} values were previously reported in literature for both mackerel and horse mackerel hydrolysates.

From the results stated above, it can be concluded that the hydrolysates of three species considered presented high values of ACE- inhibitory acitivity, the IC_{50} values ranging from 345 µg protein/mL for mackerel to 400 µg protein/mL for sardine. Nevertheless, higher values of *in vitro* ACE-inhibitory activities have been reported in literature for marine sources (Bordenave-Juchereau 2008, Ryan et al. 2011). The three species proposed are a good substrate for protein hydrolysis in order to obtain peptidic fracitions exhibiting antihypertensive activity. Nevertheless, further research is required in order to identify the active sequences responsible for the ACE inhibition (e.g., by means of size exclusion chromatography). This knowledge is essential for the subsequent fractionation and purification stages (e.g., by membrane technology), aiming to obtain a final concentrate of active peptides intended for dietary or pharmaceutical applications.

ACKNOWLEDGEMENTS

This work was supported by the Spanish National Plan I+D+i (projects CTQ2008-02978 and CTQ2011-23009). The Provincial Fisheries Development Centre from Motril (Spain) is also thank ful for their support to these research projects.

REFERENCES

A.O.A.C. (Association of Official Analytical Chemist). 2006. Official Methods of Analysis of the AOAC, Washington, USA.

Adam, A., M. Cugno, G. Molinaro, M. Perez, Y. Lepage and A. Agostoni. 2002. Aminopeptidase P in individuals with a history of angio-oedema on ACE inhibitors. Lancet. 359: 2088–2089.

Adler-Nissen, J. 1986. Enzymic hydrolysis of food proteins. London: Elsevier Applied Science Publishers, London. Great Britain.

Blanco, M., C.G. Sotelo, M.J. Chapela and R.I. Perez-Martin. 2007. Towards sustainable and efficient use of fishery resources: present and future trends. Trends Food Sci. Technol. 18: 29–36.

Bordenave, S., I. Fruitier, I. Ballandier, F. Sannier, A. Gildberg, I. Batista and J.-M. Piot. 2002. HPLC preparation of fish waste hydrolysate fractions. Effect on guinea pig ileum and ACE activity. Prep. Biochem. Biotechnol. 32: 65–77.

Bordenave-Juchereau, S. 2008. Marine by-products originating peptides that inhibit angiotensin-I-converting enzyme. *In:* J.-P. Bergé [ed.]. Added value to fisheries waste. Transworld Research Network, Kerala, India pp. 197–209.

Bougatef, A., N. Nedjar-Arroume, R. Ravallec-Plé, Y. Leroy, D. Guillochon, A. Barkia and M. Nasri. 2008. Angiotensin I-converting enzyme (ACE) inhibitory activities of sardinelle (Sardinella aurita) by-products protein hydrolysates obtained by treatment with microbial and visceral fish serine proteases. Food Chem. 111: 350–356.

Byun, H.-G. and S.-K. Kim. 2001. Purification and characterization of angiotensin I converting enzyme (ACE) inhibitory peptides from Alaska pollack (Theragra chalcogramma) skin. Process Biochem. 36: 1155–1162.

Cadée, J.A., C.Y. Chang, C.W. Chen, C.N. Huang, S.L. Chen and C.K. Wang. 2007. Bovine Casein Hydrolysate (C12 Peptide) Reduces Blood Pressure in Prehypertensive Subjects. Am. J. Hypertens. 20:1–5.

Camacho, F., P. González-Tello, M.P. Páez-Dueñas, E.M. Guadix and A. Guadix. 2001. Correlation of base consumption with the degree of hydrolysis in enzymatic protein hydrolysis. Journal of Dairy Research. 68: 251–265.

Campbell, D.J. 2003. The renin-angiotensin and the kallikrein-kinin systems. Int. J. Biochem. Cell B. 35: 784–791.

Carbonell, A., D. S. Rainieri and P. Martin. 1997. Discards of the western Mediterranean trawl fleets. Contract N° MED 94/027.

Coulter, D.M. and I.R. Edwards. 1987. Cough associated with captopril and enalapril. Brit. M. J. 294: 1521–1523.

Cushman, D.W. and H.S. Cheung. 1971. Spectrophotometric assay and properties of the angiotensin-converting enzyme of rabbit lung. Biochem. Pharmacol. 20: 1637–1648.

Ewart, S.H., A. D. Dennis, C. Barrow and A. Potwin. 2007. Anti-hypertensive dietary supplement. US Patent No. 7,179,793.

Fahmi, A., S. Morimura, H.C. Guo, T. Shigematsu, K. Kida and Y. Uemura. 2004. reduction of angiotensin I converting enzyme inhibitory peptides from sea bream scales. Proc. Biochem. 39: 1195–1200.

FitzGerald, R.J., B.A. Murray and D.J. Walsh. 2004. Hypotensive Peptides from Milk Proteins. J. Nutr. 134: 980–988.

Fitzgerald, R.J. and H. Meisel. 2000. Milk protein-derived peptide inhibitors of angiotensin-I-converting enzyme. Brit. J. Nutr. 84: S33–S37.

FROM. 2008. Asistencia técnica para la realización de una medida innovadora para la recuperación, gestión y valorización de los descartes generados por la flota española que faena en los caladeros atlántico y mediterráneo. URL: http://www.arvi.org/I+D+I/DESCARTES.pdf.

Fujita, H., K. Eiichiyokoyama and M. Yoshikawa. 2000. Classification and Antihypertensive Activity of Angiotensin I-Converting Enzyme Inhibitory Peptides Derived from Food Proteins. J. Food Sci. 65: 564–569.

Fujita, H. and M. Yoshikawa. 1999. LKPNM: A prodrug-type ACE-inhibitory peptide derived from fish protein. Immunopharmacology 44: 123–127.

Gormley, R. 2006. Fish as a functional food. Food Sci. Technol. 20: 25–28.

Hata, Y., M. Yamamoto, M. Ohni, K. Nakajima, Y. Nakamura and T. Takano. 1996. A placebo-controlled study of the effect of sour milk on blood pressure in hypertensive subjects. Am. J. Clin. Nutr. 64: 767 –771.

He, H.L., H. Wu, X.L. Chen, M. Shi, X.Y. Zhang, C.Y. Sun, Y.Z. Zhang and B.C. Zhou. 2008. Pilot and plant scaled production of ACE inhibitory hydrolysates from Acetes chinensis and its *in vivo* antihypertensive effect. Biores. Technol. 99: 5956–5959.

Hirota, T., K. Ohki, R. Kawagishi, Y. Kajimoto, S. Mizuno, Y. Nakamura and M. Kitakaze. 2007. Casein hydrolysate containing the antihypertensive tripeptides Val-Pro-Pro and Ile-Pro-Pro improves vascular endothelial function independent of blood pressure–lowering effects: contribution of the inhibitory action of Angiotensin-Converting Enzyme. Hypertens. Res. 30: 489–496.

Hiroyuki, F. 2008. Angiotensin converting enzyme inhibitor. European Patent No. 1,092,724.

Holmquist, B., P. Bünning and J.F. Riordan. 1979. A continuous spectrophotometric assay for angiotensin converting enzyme. Anal. Biochem. 95: 540–548.

Hong, F., L. Ming, S. Yi, L. Zhanxia, W. Yongquan and L. Chi. 2008. The antihypertensive effect of peptides: A novel alternative to drugs? Peptides. 29: 1062–1071.

Iroyukifujita, H., K. Eiichiyokoyama and M. Yoshikawa. 2000. Classification and antihypertensive activity of angiotensin I converting enzyme inhibitory peptides derived from food proteins. J. Food Sci. 65: 564–569.

Ishida, Y., Y. Shibata, I. Fukuhara, Y. Yano, I. Takehara and K. Kaneko. 2011. Effect of an Excess Intake of Casein Hydrolysate Containing Val-Pro-Pro and Ile-Pro-Pro in Subjects with Normal Blood Pressure, High-Normal Blood Pressure, or Mild Hypertension. Biosci. Biotech. Biochem. 75: 427–433.

Jauhiainen, T., H. Vapaatalo, T. Poussa, S. Kyrönpalo, M. Rasmussen and R. Korpela. 2005. *Lactobacillus helveticus* Fermented Milk Lowers Blood Pressure in Hypertensive Subjects in 24-h Ambulatory Blood Pressure Measurement. Am. J. Hypertens. 18: 1600–1605.

Je, J.–Y., K.-H. Lee, M.H. Lee and C.-B. Ahn. 2009. Antioxidant and antihypertensite protein hydrolysates produced from tuna liver by enzymatic hydrolysis. Food Res. Int. 42: 1266–1272.

Je, J.-Y., P.-J. Park, J.Y. Kwon and S.-K. Kim. 2004. A Novel Angiotensin I Converting Enzyme Inhibitory Peptide from Alaska Pollack (*Theragra chalcogramma*) Frame Protein Hydrolysate. J. Agric. Food Chem. 52: 7842–7845.

Johnston, C.I. 1992. Renin-angiotensin system: A dual tissue and hormonal system for cardiovascular control. J. Hypertens. 10: S13–S26.

Jung, W.-K., E. Mendis, J.-Y. Je, P.-J. Park, B.W. Son, H.C. Kim, Y.K. Choi and S.-K. Kim. 2006. Angiotensin I-converting enzyme inhibitory peptide from yellowfin sole (*Limanda aspera*) frame protein and its antihypertensive effect in spontaneously hypertensive rats. Food Chem. 94: 26–32.

Kannel, W.B. 2000. Elevated systolic blood pressure as a cardiovascular risk factor. Am. J. Cardiol. 85: 251–255.

Karaki, H., M. Kuwahara, S. Sugano, C. Doi, K. Doi, N. Matsumura and T. Shimizu. 1993. Oral administration of peptides derived from bonito bowels decreases blood pressure in spontaneously hypertensive rats by inhibiting angiotensin converting enzyme. Comp. Biochem. Physiol. C: Pharmacol. Toxicol. Endocrinol. 104: 351–353.

Kawasaki, T., C.J. Jun, Y. Fukushima, K. Kegai, E. Seki, K. Osajima, K. Itoh, T. Matsui and K. Matsumoto. 2002. Antihypertensive effect and safety evaluation of vegetable drink with peptides derived from sardine protein hydrolysates on mild hypertensive, high-normal and normal blood pressure subjects. Hukuoka acta medica 93: 208–218.

Kawasaki, T., E. Seki, K. Osajima, M. Yoshida, K. Asada, T. Matsui and Y. Osajima. 2000. Antihypertensive effect of valyl-tyrosine, a short chain peptide derived from sardine muscle hydrolyzate, on mild hypertensive subjects. J. Hum. Hypertens. 14: 519–523.

Kearney, P.M., M. Whelton, K. Reynolds, P. Muntner, P.K. Whelton and J. He. 2005. Global burden of hypertension: analysis of worldwide data. Lancet. 365: 217–223.

Knütter, I., C. Wollesky, G. Kottra, M.G. Hahn, W. Fischer, K. Zebisch, R.H.H. Neubert, H. Daniel and M. Brandsch. 2008. Transport of Angiotensin-Converting Enzyme Inhibitors by H+/Peptide Transporters Revisited. J. Pharmacol. Exp. Ther. 327: 432–441.

Kocatepe, D. and H. Turan. 2012. Proximate and fatty acid composition of some commercially important fish species from the sinop region of the Black Sea. Lipids. 47: 635–641.

Kohama, Y.,S. Matsumoto, H. Oka, T. Teramoto, M. Okabe and T. Mimura. 1998. Isolation of angiotensin-converting enzyme inhibitor from tuna muscle. Biochem. Biophys. Res. Commun. 155: 332–337.

Kohama, Y., S. Matsumoto, H. Oka, T. Teramoto, M. Okabe and T. Mimura. 1998. Isolation of angiotensin-I-converting enzyme inhibitor from tuna muscle. Biochem. Biophys. Res. Commun. 155: 332–337.

Korhonen, H. 2009. Milk-derived bioactive peptides: From science to applications. J. Func. Foods. 1: 177–187.

Lee, K.H., I.O. Kim, W.D. Yoon, J.K. Shin and H.C. An. 2007. A study on vertical distribution observation of giant jellyfish (Nemopilema nomurai) using acoustical and optical methods. J. Korean Soc. Fish Technol. 43: 355–361.

Lee, S.-H., Z.-J. Qian and S.-K. Kim. 2010. A novel angiotensin I-converting enzyme inhibitory peptide from tuna frame protein hydrolysate and its antihypertensive effect in spontaneously hypertensive rats. Food Chem. 118: 96–102.

Lei, L., H. Sun, D. Liu, L. Liu and S. Li. 2008. Transport of Val-Leu-Pro-Val-Pro in Human Intestinal Epithelial (Caco-2) Cell Monolayers. J. Agric. Food Chem. 56: 3582–3586.

Li, G.H., G.W. Le, Y.H. Shi and S. Shrestha. 2004. Angiotensin I-converting enzyme inhibitory peptides derived from food proteins and their physiological and pharmacological effects. Nutr. Res. 24: 469–486.

Li, Y., J. Zhou, K. Huang, Y. Sun and X. Zeng. 2012. Purification of a novel angiotensin I-converting enzyme (ACE) inhibitory peptide with an antihypertensive effect from loach (Misgurnus anguillicaudatus). J. Agric. Food Chem. 60: 1320–1325.

López-Fandiño, R., J. Otte and J. van Camp. 2006. Physiological, chemical and technological aspects of milk-protein-derived peptides with antihypertensive and ACE-inhibitory activity. Int. Dairy J. 16: 1277–1293.

Mancia, G., C. Medica, O. San Gerardo and G. de Backer. 2007. European Society of Hypertension (ESH) and European Society of Cardiology (ESC). Guidelines for the management of arterial hypertension. J. Hypertens. 25: 1105–1187.

Masuda, O., Y. Nakamura and T. Takano. 1996. Antihypertensive peptides are present in aorta after oral administration of sour milk containing these peptides to spontaneously hypertensive rats. J. Nutr. 126: 3063–3068.

Matsufuji, H., T. Matsui, S. Ohshige, T. Kawasaki, K. Osajima and Y. Osajima. 1995. Antihypertensive effects of angiotensin fragments in SHR. Biosci. Biotechnol. Biochem. 59: 1398–1401.

Matsufuji, H., T. Matsui, E. Seki, K. Osajima, M. Nakashima and Y. Osajima. 1994. Angiotensin I-converting enzyme inhibitory peptides in an alkaline protease hydrolyzate derived from sardine muscle. Biosci. Biotechnol. Biochem. 58: 2244–2245.

Matsui, T. and H. Matsufuji, E. Seki, K. Osajima, M. Nakashima and Y. Osajima. 1993. Inhibition of angiotensin I-converting enzyme by Bacillus licheniformis alkaline protease hydrolyzates derived from sardine muscle. Biosci. Biotechnol. Biochem. 57: 922–925.

Mizushima, S., K. Ohshige, J. Watanabe, M. Kimura, T. Kadowaki, Y. Nakamura, O. Tochikubo and H. Ueshima. 2004. Randomized controlled trial of sour milk on blood pressure in borderline hypertensive men. Am. J. Hypertens. 17: 701–706.

Mullally, M.M., H. Meisel and R.J. FitzGerald. 1997. Identification of a novel angiotensin-I converting enzyme inhibitory peptide corresponding to a tryptic fragment of bovine β-lactoglobulin. FEBS Letters. 402: 99–101.

Murray, B.A. and R.J. FitzGerald. 2007. Angiotensin converting enzyme inhibitory peptides derived from food proteins: Biochemistry, bioactivity and production. Curr. Pharm. Design. 13: 773–791.

Nagai, T., N. Suzuki, Y. Tanue, N. Kai, Y. Nakazawa and T. Nagashima. 2008. Angiotensin I-converting enzyme inhibitory activities of enzymatic hydrolysates from lees as by-

product produced during sake making using fugu muscle and fin. J. Food Agric. Environ. 6: 19–22.
Nii, Y., K. Fukuta, R. Yoshimoto, K. Sakai and T. Ogawa. 2008. Determination of antihypertensive peptides from an izumi shrimp hydrolysates. Biosci. Biotechnol. Biochem. 72: 861–864.
Ono, S., M. Hosokawa, K. Miyashita and K. Takahashi. 2003. Isolation of peptides with angiotensin I-converting enzyme inhibitory effect derived from hydrolysate of upstream chum salmon muscle. J. Food Sci. 68: 1611–1614.
Ono, S., M. Hosokawa, K. Miyashita and K. Takahashi. 2006. Inhibition properties of dipeptides from salmon muscle hydrolysate on angiotensin I-converting enzyme. Int. J. Food Sci. Technol. 41: 383–386.
Osajima, K., E. Seki and M. Yoshida. 2006. Peptide Y-2. US Patent No. 7,112,571.
Osajima, K. and Y. Ooishi. 2005. Peptide SY. US Patent No. 6,905,704.
Otani, L., T. Ninomiya, M. Murakami, K. Osajima, H. Kato and T. Murakami. 2009. Sardine peptide with angiotensin i-converting enzyme inhibitory activity improves glucose tolerance in stroke-prone spontaneously hypertensive rats. Biosci. Biotechnol. Biochem. 73: 2203–2209.
Pina, A.S. and A.C.A. Roque. 2009. Studies on the molecular recognition between bioactive peptides and angiotensin-converting enzyme. J. Mol. Recognit. 22: 162–168.
Riordan, J. 2003. Angiotensin-I-converting enzyme and its relatives. Genome Biology. 4: 225.
Roberts, P.R.,J.D. Burney, K.W. Black and G.P. Zaloga. 1999. Effect of Chain Length on Absorption of Biologically Active Peptides from the Gastrointestinal Tract. Digestion. 60: 332–337.
Ruilope, L.M. 1999. The Kidney as Part of the Cardiovascular System. J. Cardiovasc. Pharm. 33: S7–S10.
Ryan, J.T., P. R. Reynolds, D. Bolton, G.F. Fitzgerald and C. Stanton. 2011. Bioactive Peptides from Muscle Sources: Meat and Fish. Nutrients 3: 765–791.
Saito T. 2008. Antihypertensive peptides derived from bovine casein and whey proteins. *In:* Z. Bosze [ed.]. Bioactive Components of Milk. Springer, New York, USA pp. 295–317.
Saito, Y., K. Wanezaki, A. Kawato and S. Imayasu. 1994. Structure and activity of angiotensin I converting enzyme inhibitory peptides from sake and sake lees. Biosci Biotech Biochem. 58: 1767–71.
Samaranayaka, A.G.P., D.D. Kitts and E.C.Y. Li-Chan. 2010. Antioxidative and angiotensin-I-converting enzyme inhibitory potential of a Pacific hake (Merlucclus productos) fish protein hydrolysate subjected to simulated gastrointestinal digestion and caco-2 cell permeation. J. Agric. Food Chem. 58: 1535–1542.
Sánchez, D., M. Kassan, M. del M. Contreras, R. Carróna, I. Recio, M.J. Monteroa and M.A. Sevilla. 2011. Long-term intake of a milk casein hydrolysate attenuates the development of hypertension and involves cardiovascular benefits. Pharmacol. Res. 63: 398–404.
Satake, M., M. Enjoh, Y. Nakamura, T. Takano, Y. Kawamura, S. Arai and M. Shimizu. 2002. Transepithelial transport of the bioactive tripeptide, Val-Pro-Pro, in human intestinal Caco-2 cell monolayers. Biosci. Biotechnol. Biochem. 66: 378–384.
Seppo, L., T. Jauhiainen, T. Poussa and R. Korpela. A fermented milk high in bioactive peptides has a blood pressure-lowering effect in hypertensive subjects. Am. J. Clin. Nutr. 77: 326–330.
Shalaby, S.M., M. Zakora and J. Otte. 2006. Performance of Two Commonly Used Angiotensin-Converting Enzyme Inhibition Assays Using FA-PGG and HHL as Substrates. J. Dairy Res. 73: 178–186.
Sibony, M., J. Gasc, F. Soubrier, F. Alhenc-Gelas and P. Corvol. 1993. Gene expression and tissue localization of the two isoforms of angiotensin I converting enzyme. Hypertension 21: 827–835.
Sipos, T. and J. R. Merkel. 1970. An effect of calcium ions on the activity, heat stability, and structure of trypsin. Biochemistry 9: 2766–2775.

Takano, D.T. 2002. Anti-hypertensive activity of fermented dairy products containing biogenic peptides. Anton. Leeuw. Int. J. G. 82: 333–340.

Torres J.A., Y.C. Chen, J. Rodrigo-García and J. Jaczynski. 2007. Recovery of by-products from seafood processing streams. *In:* F. Shahidi [ed.]. Maximizing the Value of Marine By-products. CRC Press, Boca Raton, USA pp. 65–90.

Tuomilehto, J., J. Lindström, J. Hyyrynen, R. Korpela, M.L. Karhunen, L. Mikkola, T. Jauhiainen, L. Seppo and A. Nissinen. Effect of ingesting sour milk fermented using Lactobacillus helveticus bacteria producing tripeptides on blood pressure in subjects with mild hypertension. J. Hum. Hypertens. 18: 795–802.

Vermeirssen, V., J. Van Camp and W. Verstraete. 2002. Optimisation and validation of an angiotensin-converting enzyme inhibition assay for the screening of bioactive peptides. J. Biochem. Bioph. Meth. 51: 75–87.

Vermeirssen, V., J. Van Camp and W. Verstraete. 2004. Bioavailability of Angiotensin I Converting Enzyme Inhibitory Peptides. Brit. J. Nutr. 92: 357–366.

Wu, H., H.L. He, X.L. Chen, C.Y. Sun, Y.Z. Zhang and B.C. Zhou. 2008. Purification and identification of novel angiotensin-I-converting enzyme inhibitory peptides from shark meat hydrolysate. Process Biochem. 43: 457–461.

Wu, J., R.E. Aluko and A.D. Muir. 2002. Improved method for direct high-performance liquid chromatography assay of angiotensin-converting enzyme-catalyzed reactions. J. Chromatogr. A. 950: 125–130.

Yokoyama, K., H. Chiba and M. Yoshikawa. 1992. Peptide inhibitors for angiotensin I-converting enzyme from thermolysin digest of dried bonito. Biosci. Biotechnol. Biochem. 56: 1541–1545.

Yoon, H.-D., Y.-K. Kim, C.-W. Lim, S.-M. Yeun, M-.H. Lee, H-.S. Moon, N.-Y. Yoon, H.-Y. Park and D.-S. Lee. 2011. ACE-Inhibitory properties of proteolytic hydrolysates from giant jellyfish Nemopilema nomurai. Fish. Aquat. Sci. 14: 174–178.

5

Antimicrobial Peptides from Fish

Florie Desriac, Camille Jégou, Benjamin Brillet, Patrick Le Chevalier and *Yannick Fleury**

ABSTRACT

Antimicrobial peptides (AMPs) are ubiquitous in the living kingdom. Some bioactive peptides have shown numerous bioactivities such as antibacterial, antifungal and antiviral. In this chapter, the general properties of antimicrobial peptides from fish are discussed. Moreover, the authors have focussed on AMPs known to date, produced by farmed fish and fish-associated bacteria. An emerging category of AMPs-like molecules consisting of proteolytic fragments of inactive protein is also discussed. Microbial infections are among the most destructive diseases that affect vertebrate and invertebrate species in aquaculture. Despite the impact that diseases have on aquatic organisms, we know relatively little about what farmers can do to prevent and treat infections and how fight fish diseases. Difficulties in controlling infections in aquaculture are mainly due to the absence of commercial vaccines and the absence of specific therapeutic agents. In the long term, alternative treatments using antimicrobial peptides and probiotics may be developed.

Laboratoire Universitaire de Biodiversité et d'Ecologie Microbienne, Université de Bretagne Occidentale, 29000 Quimper, France.

Emails: desriac@univ-brest.fr; cjegou@univ-brest.fr; Benjamin.Brillet@univ-brest.fr; lecheval@univ-brest.fr

* Corresponding author: fleury@univ-brest.fr

INTRODUCTION

The ocean represents more than two-thirds of the earth's surface. The marine environment presents one of the greatest biodiversities on earth at both the eukaryotic and prokaryotic levels. Indeed, marine organisms live in a real microbial "soap". Bacterial and fungal concentrations have been estimated to be around respectively 10^4 CFU.mL^{-1} and 10^3 CFU.mL^{-1} (Zweifel and Hagstrom 1995) while viral particle concentration is evaluated at 3.10^6 VLP. mL^{-1} (Suttle 2005). Beyond this dense biodiversity, the liquid nature itself of the aquatic environment introduces proximity and continuity. To survive and thrive in such a medium, aquatic animals have had no alternative than to associate with beneficial micro-organisms to occupy the ecological niche and to prevent pathogenic implantation and/or to produce potent chemical weapons to fight against the pathogenic microbes. In the light of such selection pressure, marine animals may constitute a huge reservoir of efficient antimicrobial compounds and of symbiotic association with microorganisms. Paradoxically, antimicrobial compounds from marine sources, are anecdotic compared to the terrestrial sources, probably due to technical difficulties involved in sample collection. Investigation of marine pharmacology literature for antimicrobial compounds from 1998 to 2010 resulted in the identification of four main structural families—polyketides (27%), terpenoids (20%), peptides (16%) and alkaloids (15%) (Mayer and Hamann 2002, 2004, 2005, Mayer et al. 2007, 2009, 2011). To date, various antimicrobial compounds from the sea have been isolated, characterized and investigated in clinical trials (for review see references Donia and Hamann 2003, Blunt et al. 2004, Bhatnagar and Kim 2010, Montaser and Luesch 2011).

Antimicrobial peptides (AMPs) have been extensively studied in the terrestrial living kingdom (for review see references Andreu and Rivas 1998, Hancock and Scott 2000, Zasloff 2002, Tennessen 2005). These gene-encoded peptides have been found in most branches of the tree of life. Nowadays, AMPs are plethoric and represent a hyphen between innate and adaptative immunity (Lawniczak et al. 2007, Iwanaga and Lee 2005, Kurtz and Armitage 2006, Patrzykat and Douglas 2005). Indeed, AMPs are the main component of biochemical defence systems in invertebrates while these peptides ensure a first line of defence in vertebrates, complementing adaptative immunity (Hancock and Scott 2000, Wiesner and Vilcinskas 2010). Because of the multitude of AMPs described among living organisms, providing an exhaustive review is a difficult task, and is practically unworkable due to constant literature publications (around 780 publications with AMP in the title or abstract in 2011 using the Pubmed database). A large number of excellent reviews regarding AMPs have been published according to the producer organisms. Various global or specialized databases have been built (Table 1). The very vast application potentialities of AMPs have aroused a keen interest in these molecules. They have been described as effective therapeutants to prevent or fight infections in humans (Hancock and Chapple 1999, Hancock and Patrzykat 2002, Reddy et al. 2004, Hancock and Sahl 2006, Zaiou 2007). Various medical applications have been proposed such as dental (Miyasaki and Lehrer 1998) and skin (Ulvatne 2003) treatments. Other applications have also been suggested, for instance, food preservatives (Devlieghere et al. 2004, Rydlo et al. 2006). Few applications in animal breeding have also been evoked. Animal breeding of course implies animal concentration and so infections are the most destructive. Despite their impact, few solutions exist. In the long term, alternative treatments using antiviral and antibacterial drugs may be developed in the the most effective way for sustainable breeding production.

In the first part of this review, the general properties of antimicrobial peptides are described. However, the purpose of this review is to study marine AMPs known to date produced by fish species. We have also focused on an emerging category of AMPs that can be described as antimicrobial secondary peptides. Indeed, since a few years, certain AMPs characterized from invertebrate and vertebrate organisms have appeared as proteolytic fragments of inactive protein involved in other biological function(s), e.g., structural proteins. In so far as they result from *in vivo* enzymatic cleavage of an inactive protein, we have chosen to name this kind of molecule "antimicrobial post-peptide". Particular attention will also be paid to microbial association in fish and the third part of this review will be dedicated to antimicrobial activities produced by fish-associated bacteria.

Table 1 A selection of databases dedicated to AMPs.

Database		Link	Total peptides
APD2	generalist	http://aps.unmc.edu/AP/main.php	1997
cAMP	generalist	http://www.bicnirrh.res.in/antimicrobial/	4020
DAMPD	generalist	http://apps.sanbi.ac.za/dampd/	1232
Bactibase	bacteriocin	http://bactibase.pfba-lab-tun.org/main.php	177
CyBase	cyclic peptide	http://www.cybase.org.au/	633
Defensin Knowledgebase	defensin	http://defensins.bii.a-star.edu.sg/	363
PhyAMP	plant AMP	http://phytamp.pfba-lab-tun.org/main.php	271

HISTORY OF AMP

The oldest and perhaps the most well known antimicrobial peptide, nisin, was first described in 1928. This 34-residue peptide is produced by a strain of lactic acid bacteria, *Lactococcus lactis*. It belongs to the lantibiotics, a class of highly post-translationally modified bacteriocins. Its structure was elucidated in 1971 (Gross and Morell 1971). Nisin contains unusual amino acids (unsaturated and stereoinverted amino acids) as well as thioether bridges.

Another ancient and nonetheless well known AMP is melittin (Fischer and Neumann 1961). Unlike nisin, melittin is a 26 non-modified residue peptide. Melittin is the main peptide component of the venom of the European honeybee, *Apis mellifera*. These two ancient peptides represent the great diversity of AMPs. Indeed, these gene-encoded peptides are light-years away at structural and activity levels. In fact, the first, highly modified, presents a narrow spectrum of activity while the second, unmodified, is active against prokaryotic and eukaryotic cells.

Until the eighties, these two AMPs did not arouse great interest within the scientific community. This class of antibiotics was re-discovered at the end of this decade with widespread antibiotics bioprospecting giving birth to a "second generation" of AMPs. Cecropin from an insect in 1981 and then magainin from an amphibian in 1987 were characterized (Steiner et al. 1981, Zasloff 1987). Magainin was the first AMP from a vertebrate and its discovery struck a chord in the scientific community. These two linear peptides aroused keen interest because AMPs appeared for the first time as potent weapons to fight pathogenic bacteria. AMPs were then extensively researched and publications dedicated to AMPs grew exponentially. They were purified and characterized from plants (García-Olmedo et al. 1998), insects (Otvos 2000), amphibians (Rinaldi 2002), birds (Sugiarto and Yu 2004) and mammals (Pazgier et al. 2006). To date, thousands of structures have been documented from eukaryotic and prokaryotic organisms. Various increasingly online databases are available. Some are dedicated to AMPs in general, grouping thousands of peptides experimentally validated or predicated while other databases are devoted to specific AMPs such as cyclic or bacterial AMPs (Table 1). Ribosomally synthesized AMPs have appeared as excellent candidates for therapeutic and alternative preservatives giving birth to this success story.

GENERAL PROPERTIES OF AMPs

AMPs have been shown to be ubiquitous and to be produced by organisms ranging from bacteria to superior mammals. Ribosomally synthetized AMPs from prokaryotes are called bacteriocins. They have been demonstrated to be potent weapons to defend or to capture an ecological niche (Riley and Wertz 2002, Dobson et al. 2012). In Eukaryotes, antimicrobial peptides like thionins have been described in various organs of plants (for review see García-Olmedo et al. 1998, Pelegrini and Franco 2005, Stec 2006). Nevertheless most AMPs have been investigated in animals. In insects and amphibians, a multitude of AMPs have been described. They are also produced by birds and mammals (for review see references Sugiarto and Yu 2004, Pazgier et al. 2006). In a nutshell, during the last quarter of a century, a plethora of ribosomally synthetized AMPs have been described and characterized. From bacteria to superior mammals, their important role in the organism's defence has been proven.

In spite of their great diversity of origins, AMPs show certain common structural features. Indeed, most AMPs are 20- to 60-residue long, cationic and amphipathic peptides. Based on a 1997 peptide sequence database, statistical data reveal that the average AMP is 30.5 residues long and exhibits a cationic net charge, +3.1 (APD2, see Table 1). Of course, exceptions exist in terms of size or charge (Wang et al. 2003, Dennison et al. 2006, Sitaram 2006, Wang et al. 2007). Overall, two main categories can be distinguished, i.e., linear peptides and disulfide bridge-containing peptides. A global classification based on structural homology is now accepted. Four classes of gene-encoded AMPs have been defined. The first contains linear peptides with an amphipathic α-helical conformation, the second brings together linear peptides with a great proportion of one residue such as P, R or W, the third is composed of peptides containing loop structures while β-strand structures containing disulfide bridge(s) form the fourth class.

AMP activities against viruses, bacteria, fungi and parasitic cells have been described. However, according to the AMP, a great disparity in the spectrum of activity does exist. Indeed, AMPs can demonstrate either a very large spectrum, being active against prokaryotic and eukaryotic cells, or a very narrow one, with antimicrobial activity limited to only a few bacterial genera. It is now well established that the primary target of these peptides is the plasmatic membrane. They act on the sensitive cells by disrupting the plasmatic membrane (Epand and Vogel 1999, Shai 2002, Yeaman and Yount 2003). Several mechanistic models (barrel-stave model, carpet-like model, torroid pores model) have been proposed (Fig. 1). Metabolic targets have also been described for different peptides, e.g., protein synthesis inhibition, nucleic acid binding, cell-wall biosynthesis inhibition, etc. (for review see Yeaman and Yount 2003, Brogden 2005, Jenssen et al. 2006, Nguyen et al. 2011). The mechanism leading to cellular death may ensue from membrane and metabolic interactions. The balance between these two modes of action may depend on the peptide concentration. Their selectivity of activity and mechanism of action make AMPs an excellent model to study peptide/membrane interactions. Because many peptides act by perturbing membranes, their ability to inactivate enveloped viruses has been studied (Yasin et al. 2000). AMPs are more effective on enveloped viruses than non-enveloped viruses, and may act at different stages during the course of viral infection to inhibit viral replication (Daher et al. 1986). Three mechanisms have been proposed to account for antiviral action: (1) direct inactivation of viral particles by perturbing lipid bilayers of viral envelopes (Daher et

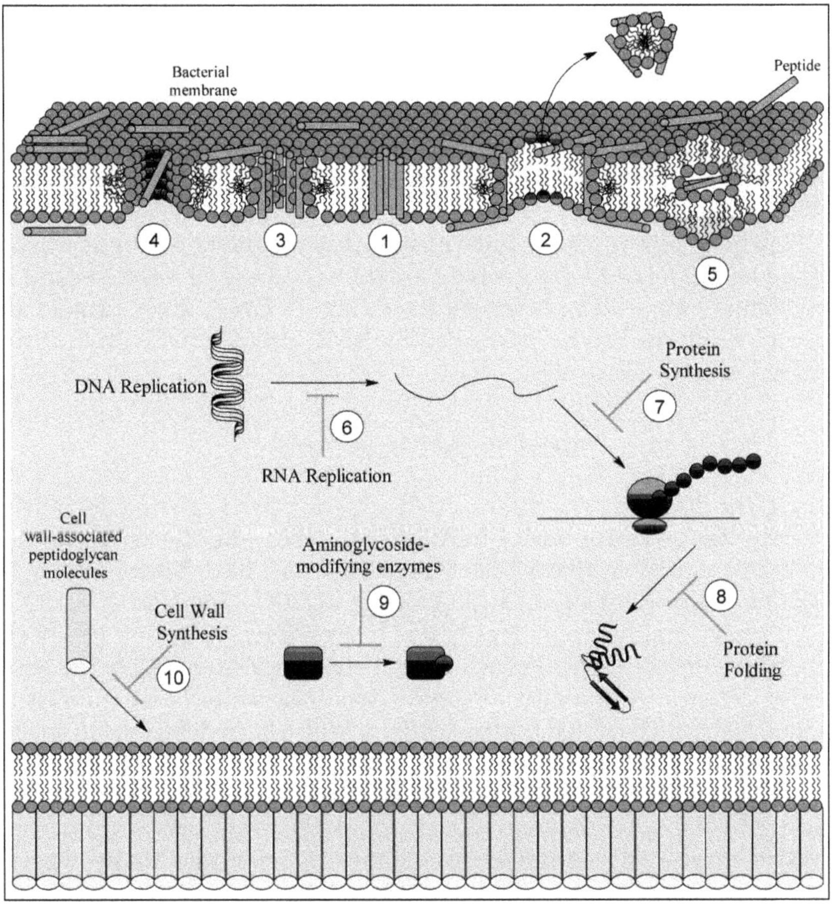

Fig. 1 Common mechanisms of action of antimicrobial peptides on bacteria, declined in membrane permeabilization processes (1 to 5) and intracellular effects (6 to 10) (adapted from Jenssen et al. 2006 and Nguyen et al. 2011). When reaching a threshold concentration, the peptides at the outer surface of the membrane can insert themselves into the membrane and then (1) form peptide-lined pores according to the barrel-stave model, (2) create micellar structures with the membrane by interacting with its lipids, in the carpet model, or (3) form peptide-and-lipid-lines pores according to the toroidal pore model. In comparison, the discordered toroidal model involves less peptides (4). The coming of peptides in between the membrane can increase its thickness, sometimes generating a non-bilayer intermediate form (5). All these mechanisms of action can occur altogether for the same molecule. In the cytoplasm, various disturbances can be generated by AMPs, from DNA replication to cell-wall synthesis. Buforin II, pleurocidin and dermaseptin can inhibit DNA and RNA synthesis (6). Indolicidin and PR-39 are known to limit protein synthesis in targeted bacterial cells (7). Other kinds of enzymatic activity can also be altered by AMPs. Pyrrhocidin limits the ATPase activity DnaK, which is implied in the chaperone-assisted protein folding (8). The enzymatic modification of aminoglycosydes has also been shown to be inhibited by AMPs (9). Finally, lantibiotics such as the well-known niscin and mersacidin have been demonstrated to disturb the synthesis of the cell-wall molecules (10), among other structural components.

al. 1986); (2) prevention of viral penetration into the host cell by inhibiting viral-cellular membrane fusion (Baghian et al. 1997); and (3) inhibition of viral replication in infected cells by suppressing viral gene expression (Wachinger et al. 1998). Studies have shown that the same antimicrobial peptide may act on different viruses through different mechanisms (Daher et al. 1986, Aboudy et al. 1994, Marcos et al. 1995, Baghian et al. 1997, Wachinger et al. 1998).

However, recent results have shown that the membranolytic activity may be comparable to the visible part of the iceberg. Indeed, secondary intracellular targets have been identified (Fig. 1). DNA, RNA, protein and cell-wall synthesis can be perturbed by AMP interactions with enzymes involved or with substrates.

MARINE AMPs (mAMPs)

During the past twenty years, AMPs have been reported in marine invertebrates (Arthropoda, Chordata, Mollusca and Gastropoda) and vertebrates mostly in teleost fish (Patrzykat et al. 2003, Tincu and Taylor 2004, Otero-González et al. 2010, Smith et al. 2010, Sperstad et al. 2011). mAMPs have mainly been investigated in economically important species with the praiseworthy goal of achieving better understanding of the host's natural defence. Aquacultural species have therefore been extensively studied. It should be noted that mAMPs have also been described in species of minor economical interest such as *Aurelia* sp. (Cnidaria) (Ovchinnikova et al. 2006), *Limulus* sp. (Chelicerata) (Iwanaga 2002), *Styela* sp. (Urochordata) (Tincu et al. 2003), *Dolabella* sp. (Gastropoda) (Iijima et al. 2003b), etc. These peptides are at least as attractive as the former, being potentially effective alternative therapeutants. In this review however, we chose to focus exclusively on mAMPs produced by fish species. For more information, the reader is referred to recent reviews (Patrzykat and Douglas 2003, Tincu and Taylor 2004). We also decided not to compare the antimicrobial activity of these peptides. Indeed, it should be noted that it is risky to analyse the antibacterial and/or antifungal spectrum of activity of AMPs because of the absence of a standardized assay. There is a great diversity of protocols to estimate antimicrobial activity (various target cell concentrations (10^4 to 10^8 UFC.mL^{-1}), physiological state, strains, incubation time or temperature, culture medium, liquid or diffusion assay, etc.). All these parameters may influence antimicrobial activity at different levels and therefore impede rigorous comparison.

FISH AMPs

Unlike invertebrates which depend on innate effectors for warding off microbial invasion, adaptative immunity is present in vertebrates. In fish in particular, the acquired immune response is apathetic due to a limited antibody panel and slow lymphocyte proliferation (Magnadottir 2006). At least, adaptative immunity in fish has the merit of existing. So AMPs, one of the main components of fish innate immunity, play a crucial role in fish defence. Moreover, various peptides have been characterized (Table 2)

Table 2 Overview of antimicrobial peptides from fish known to date.

Fish	Name	Length	Activity
Anguilla japonica	Ajl-2	142	Antibacterial
Ctenopharyngodon idella	LEAP	78	Antimicrobial
Danio rerio	hepcidin 1	20	Antimicrobial
Dicentrarchus labrax	Dicentracin	22	Antimicrobial
Epinephelus coioides	Epinecidin	25	Antimicrobial
	Hepcidin		
Glyptocephalus cynoglossus	Pleurocidin-like peptides	26	Antimicrobial
Grammistes sexlineatus	Grammistins	26	Antibacterial
Hippoglossoides platessoides	Pleurocidin-like peptides	26	Antimicrobial
	Hipposin	51	Antibacterial
Ictalurus punctatus	histone-like protein	20	Antimicrobial
	LEAP	73	Antimicrobial
Lateolabrax japonicus	Hepcidin-like peptide	21	Antimicrobial
Limanda ferruginea	Pleurocidin-like peptides	25	Antimicrobial
Misgurnus anguillicaudatus	Misgurin	21	Antimicrobial
Morone chrysops	Moronecidins	23	Antimicrobial
Morone chrysops x Morone saxatilis	Hepcidin	21	Antibacterial
	Moronecidins	22	Antimicrobia
Morone saxatilis	Moronecidins	23	Antibacterial
Myxine glutinosa	HAP	38	Antimicrobia
Oncorhynchus mykiss	Salmocidins	13	Antibacterial
	Oncorhyncins	69	Antibacterial
	Histone H2A	12	Antimicrobia
	β-defensin 1	60	Antibacterial
	Hepcidin	25	Antibacterial
	LEAP2B	77	Antimicrobial
Pagrus major	Chrysophsins		
	hepcidin	85	Antimicrobial
Pardachirus marmoratus	Pardaxin P	33	Antimicrobial
Pelteobagrus fulvidraco	LEAP	78	Antimicrobial
Petromyzon marinus	CRP	19	Antimicrobial
	Pleurocidin-like peptides	21	Antimicrobial
Salmo salar	SAMPH1	30	Antibacterial
Silurus asotus	Parasin I	19	Antibacterial
Siniperca chuatsi	Moronecidin	22	Antibacterial

Abbreviations: LEAP: liver-expressed antimicrobial peptide, HAP: hematopoietic antimicrobial peptide, CRP: Corticostatin-related peptide. Antimicrobial differs from Antibacterial by fungal activity.

such as pardaxin from *Pardachirus* sp. (Porcelli et al. 2004), Misgurnin from *Misgurnus* sp. (Park et al. 1997) and Moronecidin from *Morone* sp. (Lauth et al. 2002, Campagna et al. 2007, Salerno et al. 2007). In this review, we will focus on mAMPs found in the main marine fish-farmed species, namely Salmononids (e.g., *Salmo salar*), Moronids (e.g., *Dicentrarchus labrax*) and Scophthalmid (e.g., *Scophthalmus maximus*). Based on both structural features and origins, we have classified fish AMPs into four classes:

(i) AMPs exhibiting an α-helical structure (pleurocidins, moronecidins and piscidins)
(ii) β-sheet structured AMPs containing 4 disulfide bonds (hepcidins)
(iii) proteolytic fragments of structural or functional proteins
(iv) AMPs produced by fish-associated bacteria

Linear Helical AMPs from Fish

This family of AMPs brings together AMP families such as pleurocidins from flatfish, grammistins from soapfish, moronecidins and piscidins from striped bass, chrysphosins from sea bream and dicentracin from sea bass, gaducidins from atlantic cod, and epinecidin from grouper. These 13- to 28-residue long linear peptides adopt an α-helical structure in membrane environments (Table 3).

Grammistins

Grammistins have been classified into three sub-groups according to sequence homologies (Kaji et al. 2006). Sub-groups 1 and 3 gathered 24 to 28-residue long peptides while shorter peptides, 13-residue long, made up the second sub-group 2 (Table 3). These peptides exhibit antimicrobial activity against Gram-positive and Gram-negative bacteria at μM levels. They have been shown to exercise their antimicrobial activity in a lytic way after binding to membrane phospholipids (Sugiyama et al. 2005). All grammistins are synthetized as a long precursor made of a highly conserved 21-residue long signal peptide and a 53 to 63 propeptide containing an acidic propiece and a unique copy of grammistin (Kaji et al. 2006). Such an organization has already been described for pleurocidins and dermaseptins (see below).

Chrysophsins

Chrysophsins are 19- to 25-residue long peptides purified from sea-bream gills (Iijima et al. 2003a) (Table 3). Chrysophsins have mainly been studied at the levels of activity and mechanism of action. Chrysophsin-1 and -3

Table 3 Linear helical peptides isolated from fish.

Name	Producer	isolated from	Sequence	Activity	References
Pleurocidins					
WF1	*P. americanus*	skin-secreted mucous	GWGSFFKKAAHVGKHVGKAALTHYL	Gram+, Gram-,	(Cole et al. 1997)
WF2	*P. americanus*	skin-secreted mucous	GWGSFFKKAAHVGKHVGKAALTHYLG	Gram+, Gram-/ Fungus	(Douglas et al. 2003b)
WF1L	*Pp. americanus*	skin-secreted mucous	GKGRWLDRIGKAGGIIGGALDHL	Gram+, Gram-/ Fungus	(Douglas et al. 2003b)
WFX	*Pp. americanus*	skin-secreted mucous	RSTEDIIKSISGGGFLNAMNA	Gram+, Gram-/ Fungus	(Douglas et al. 2003b)
WFY	*Pp. americanus*	skin-secreted mucous	FLGFLFHGIRHGIKAIHGMIH	Gram+, Gram-/ Fungus	(Douglas et al. 2003b)
WF3	*P. americanus*	skin-secreted mucous	FLGALIKGAIHGGRFIHGMIQNH	Gram+, Gram-/ Fungus	(Douglas et al. 2001)
WF4	*P. americanus*	skin-secreted mucous	GWGSIFKHGRHAAKHIGHAAVNHYL	Gram+, Gram-/ Fungus	(Douglas et al. 2001)
Moronecidins					
Moronecidin	*Morone chrysops*	gills and skin	FFHHIFRGIVHVGKTIHKLVTG	Gram+, Gram-/ Fungus	(Lauth et al. 2002)
Moronecidin	*Morone saxatilis*	gills and skin	FFHHIFRGIVHVGKTIHRLVTG	Gram+, Gram-/ Fungus	(Lauth et al. 2002)
Dicentracin	*Dicentrarchus labrax*	cDNA Library	FFHHIFRGIVHVGKSIHKLVTG	undetermined	(Salerno et al. 2007)
Piscidins					
Piscidin-1	*Morone saxatilis*	gills and skin	FFHHIFRGIVHVGKTIHRLVTG	Gram+, Gram-/ Fungus	(Lauth et al. 2002)
Piscidin-2	*Morone saxatilis*	gills and skin	FFHHIFRGIVHVGKTIHKLVTG	Gram+, Gram-/Fungus/Viruses	(Campagna et al. 2007)
Piscidin-3	*Morone chrysops x Morone saxatilis*	mast cells in gill, skin and gut and in lining blood vessels in the viscera	FIHHIFRGIVHAGRSIGRFLTG	Gram+, Gram-	(Silphaduang and Noga 2001)
Epinecidin					
Epinecidin-1	*Epinephelus coioides*	head, kidney, intestines, skins	FIFHIIKGLFHAGKMIHGLVTRRRH	Gram+,Gram-	(Pan et al. 2007)
Grammistins					
Grammistin Pp2b	*Pogonoperca punctata*	skin secretion	FIGGIISFIKKLF	Gram+, Gram-/ hemolytic	(Sugiyama et al. 2005)
Grammistin Pp2a	*Pogonoperca punctata*	cDNA Library	FIGGIISLIKKLF	Gram+, Gram-,	(Kaji et al. 2006)
Grammistin Pp4b	*Pogonoperca punctata*	cDNA Library	LFGFLIPLLPHLIGAIPQVIGAIR	Gram+,Gram-/ hemolytic	(Kaji et al. 2006)
Grammistin Pp4a	*Pogonoperca punctata*	cDNA Library	LFGFLIPLLPHIIGAIPQVIGAIR	Gram+,Gram-/ hemolytic	(Kaji et al. 2006)
Grammistin Pp3	*Pogonoperca punctata*	skin secretion	NWRKILGQIASVGAGLLGSLLAGYE	Gram+,Gram-/ hemolytic	(Sugiyama et al. 2005)
Grammistin Pp1	*Pogonoperca punctata*	skin secretion	FIGGIISFFKRLF	Gram+,Gram-/ hemolytic	(Sugiyama et al. 2005)
Grammistin GsG	*Grammistes sexlineatus*	skin secretion	LFGFLIPLLPHIIGAIPQVIGAIR	Gram+,Gram-/ hemolytic	(Sugiyama et al. 2005)
Grammistin GsF	*Grammistes sexlineatus*	skin secretion	LFGFLIKLIPSLFGALSNIGRNRNQ	Gram+,Gram-/ hemolytic	(Sugiyama et al. 2005)
GrammistinGsC	*Grammistes sexlineatus*	skin secretion	NWRKILGKIAKVAAGLLGSMLAGYQV	Gram+,Gram-	(Sugiyama et al. 2005)
Grammistin GsA	*Grammistes sexlineatus*	skin secretion	WWRELLKKLAFTAAGHLGSVLAAKQSGW	Gram+,Gram-	(Sugiyama et al. 2005)
GrammistinGsE	*Grammistes sexlineatus*	cDNA Library	FIGGIISFIKKLF	Gram+,Gram-/ hemolytic	(Kaji et al. 2006)
Grammistin Gs D	*Grammistes sexlineatus*	cDNA Library	FIGGIISFFKRLF	Gram+,Gram-/ hemolytic	(Kaji et al. 2006)
Grammistin Gs B	*Grammistes sexlineatus*	skin secretion	IGGIISFFKRLF	Gram+,Gram-/ hemolytic	(Sugiyama et al. 2005)
Chrysophsins					
Chrysophsin-1	*Pagrus major*	gills	FFGWLIKGAIHAGKAIHGLIHRRRH	Gram+,Gram-	(Iijima et al. 2003a)
Chrysophsin-2	*Pagrus major*	gills	FFGWLIRGAIHAGKAIHGLIHRRRH	Gram+,Gram-	(Iijima et al. 2003a)
Chrysophsin-3	*Pagrus major*	gills	FIGLLISAGKAIHDLIRRRH	Gram+,Gram-	(Iijima et al. 2003a)

P. and *Pp.* respectively mean *Pleuronectes* and *Pseudopleuronectes* genera.

interact and disturb the membrane bilayer (Mason et al. 2007, Wang et al. 2011). Moreover structure/activity relationship studies have demonstrated the crucial role of the original C-terminal part RRRH in peptide insertion into lipid membrane (Mason et al. 2007).

Pleurocidins

To date about twenty different structures have been elucidated using purification protocols or a cDNA library (Patrzykat and Douglas 2003, Patrzykat et al. 2003) (Table 3). The first of these was isolated in 1997 in the skin-secreted mucous of the winter flounder *Pleuronectes americanus* (Cole et al. 1997). Since then, they have been described in various flatfish (plaice, halibut and flounder (Patrzykat et al. 2003)). These peptides exhibit a highly cationic net charge with a pI close to 10 (Cole et al. 1997, Patrzykat et al. 2003, Douglas et al. 2003b). Based on circular dichroism (CD) spectra in TFE, an α-helical peptide organization has been suggested (Cole et al. 1997). This has been confirmed using nuclear magnetic resonance (NMR) (Syvitski et al. 2005). Pleurocidin did not adopt any organized conformation in aqueous solution but in the presence of TFE even at low concentrations or in a membrane-mimicking environment such as dodecylphosphocholine micelles, it assumes an integral and rigid α-helical structure (Syvitski et al. 2005). When plotted as an α-helical wheel, the α-helix exhibited clearly distinguishable hydrophobic and hydrophilic domains. Indeed, hydrophilic and hydrophobic residues are respectively clustered into two distinct sectors. The former, located on one side of the cylindrical surface, forms a polar face of the helix with an average angle of about 60° perpendicular to the long axis of the helix. This amphipathic helical structure is classic in terrestrial vertebrate AMPs. Moreover, pleurocidins show a significant homology (up to 68% identity) with another AMP family from vertebrates, the dermaseptins, which have been isolated from the skin of an amphibian, *Phylomedusa* sp. (Cole et al. 1997, Charpentier et al. 1998).

Pleurocidins are highly potent AMPs with a wide spectrum of activity. They exhibit antibacterial activity against both gram-positive and gram-negative bacteria with an average MIC of around 50 $\mu g.mL^{-1}$ (Cole et al. 1997, Cole et al. 2000, Douglas et al. 2003b). Their antifungal activity has also been demonstrated (Jung et al. 2007). The pleurocidin mode of action was studied using a planar lipid bilayer. When the planar lipid bilayer contains anionic lipid-like phosphatidylserine it makes a well-defined ion channel; otherwise no pore-forming activity was detected. Moreover, it seems that the pleurocidin may fulfil its antimicrobial activity according to the torroidal pore model rather than the barrel-stave model (Saint et al. 2002).

Pleurocidins have been shown to be synthesized as about 70-residue long prepropeptides consisting of an N-terminal signal peptide, the active sequence and then an acidic propiece at the C-terminus (Cole et al. 2000, Douglas et al. 2003b). This precursor organization has also been described in the dermaseptin family (Amiche et al. 1999). It has been postulated that the anionic propiece interacted with the active sequence leading to an electrostatic neutralization (Cole et al. 2000). This hypothesis has been already ventured for other AMPs such as dermaseptins or mussel defensins.

Pleurocidins have been isolated from skin mucous secretions of flatfish. Using immunogold labelling and transmission electronic microscopy, pleurocidin WF2 has been localized in the mucin granule of the skin and in the intestinal gobelet cells of the flounder (Cole et al. 2000). Using a genomic library, the expression of the different pleurocidins has been shown to be tissue-specific and developmental stage-specific (Douglas et al. 2003a). It appears that this family of peptides may play an important role not only at the mucosal immunity level but also in the global organism defence system. In our knowledge, to date, no information is available regarding the bacterial challenge effect on gene regulation.

Moronecidins–Piscidins–Gaducidins

Moronecidins and piscidins have been described virtually at the same time from hybrid striped bass (Silphaduang and Noga 2001, Lauth et al. 2002). Since then, these two names have survived for the same family of AMPs. Piscidins, 3 isoforms described till date, have been isolated from mast cells, which had never been described for any animal (Silphaduang and Noga 2001). Moronecidins, two known isoforms, have been isolated from the skin and gills of sea bass (Lauth et al. 2002). To simplify, using primers designed for conserved sequence, a third isoform type named "dicentracin" was identified in leukocytes of *Dicentrarchus labrax* (Salerno et al. 2007). Recently, gaduscidins were identified using expressed sequence tags (ESTs) from head kidney, blood, brain, gill, pyloric caecum, and the spleen of non-stressed juvenile cod (Browne et al. 2011).

Howsoever that may be, all these peptides, except piscidin-4 which is a 44-residue long peptide (Park et al. 2011), are made of 22 amino acid residues. They display sequence homologies with pleurocidins (Lauth et al. 2002). Moronecidins and piscidins share the same structural features as pleurocidins, adopting an amphipathic α-helical structure (Table 3). Moronecidins and piscidins display a wide spectrum of activity, being a potent antibacterial and antifungal peptide. Piscidins have also demonstrated antiviral activity (Chinchar et al. 2004). They have also displayed haemolytic activity against human and sheep erythrocytes for concentrations close to MIC (Lauth et al. 2002). Nevertheless, moronecidins and piscidins differ from pleurocidin by a carboxy-amidation, which has never been described for any peptide in fish.

The structure-activity relationship of piscidins has been extensively investigated (Campagna et al. 2007, Lee et al. 2007, Moon et al. 2007, Park et al. 2011). Moon et al. have shown that backbone flexibility may influence antimicrobial and hemolytic activity (Moon et al. 2007). The ultimate goal is to define the structural features responsible for the selectivity and specificity of antimicrobial activity. These studies allowed for the clarification of the piscidin mode of action (Chekmenev et al. 2006, Campagna et al. 2007). Chekmenev et al. have shown using solid state NMR that piscidin-1 and -3, carboxy amidated or not, settle down in a parallel manner to the membrane surface. Like pleurocidins, using the same technology, i.e., the planar lipid bilayer technology, it appears that piscidins act on the target membrane by the torroidal pore model rather than the barrel-stave model (Campagna et al. 2007).

Piscidins, moronecidins, dicentracin, grammistins and pleurocidins precursors (79 residues long) are overall organized identically (Salerno et al. 2007). Piscidins, moronecidins are constitutively produced in multiple tissues such as gills, intestine, spleen, anterior kidney and blood (Silphaduang and Noga 2001, Lauth et al. 2002). Despite minimal differences, it appears that pleurocidins, piscidins and moronecidins belong to the same family.

The hidden part of the fish AMP activity is currently coming to light. Indeed, the ubiquitous presence of piscidin-1 in fish even in non-immune tissue strongly suggests another role(s) over and above acting as a simple effector (Ruangsri et al. 2012). Moreover piscidins have been involved in apototosis and necrotic activity in tumor cells (Lin et al. 2012). Epinecidin was shown to modulate the interleukin-10, -1b, tumor necrosis factor-α, and interferon-γ expressions in zebra fish (Pan et al. 2011). Moreover, these peptides have been assayed against various cellular targets for different applications such as anti-*Candida*, anticaryogenic agent and anti-cancer cells for pleurocidins (Cho and Lee 2011, Hilchie et al. 2011, Mai et al. 2011) showing multiple AMP applications.

Cyclic Cystein-rich AMPs from Fish

Hepcidins, also called Liver Expressed Antimicrobial Peptides (LEAP), are 20- to 26-residue long cystein-rich AMPs. They form a family of about 30 peptides distributed among vertebrates from fish to superior mammals including amphibians (Shi and Camus 2006). Hepcidins have been isolated from skin and gills in *Salmo salar* (Douglas et al. 2003a) but most of them have been predicted from a cDNA library or expressed sequence tag database (EST). They possess a typical 8-cysteine signature (CxxCxCCxxxx(x) CxxCC) engaged in four disulfide bonds (Table 4). Despite great sequence

Table 4 Cyclic Cystein-Rich Peptides: Hepcidins.

Hepcidin Producer	Sequence	Reference
Danio rerio	LCRFCCKCCRNKGCGYCCKF	(Shike et al. 2004)
	QSHLSLCRFCCKCCRNKGCGYCCKF	(Shike et al. 2004)
Hippoglossus hippoglossus	GFKCKFCCGCCRPGVCGLCCRF	(Park et al. 2005)
Ictalurus punctatus	QSHLSLCRYCCNCCKNKGCGFCCRF	(Bao et al. 2005)
Lateolabrax japonicus	AIKCKFCCGCCTPGVCGVCCRF	(Ren et al. 2006)
Lateolabrax japonicus	DCRFCCGCCTDVSGCGVCCRF	(Ren et al. 2006)
Morone chrysops x Morone saxatilis	GCRFCCNCCPNMSGCGVCCRF	(Lauth et al. 2005)
Oncorhynchus mykiss	QSHLSLCRWCCNCCHNKGCGFCCKF	(Zhang et al. 2004b)
Pseudopleuronectes americanus	GFKCKFCCGCCRAGVCGLCCKF	(Douglas et al. 2003a)
Salmo salar	QSHLSLCRWCCNCCHNKGCGFCCKF	(Douglas et al. 2003a)
	QIHLSLCGLCCNCCHNIGCGFCCK	(Douglas et al. 2003a)

homologies, hepcidins can be categorized into four classes. Differing from human hepcidins by amino terminal truncations, fish hepcidins belong to two classes (Shi and Camus 2006). Mainly using EST, fish hepcidins have been identified in Salmonids (salmon and trout, (Douglas et al. 2003a)), Moronids (sea bass) (Shike et al. 2002, Lauth et al. 2005), Serranids (grouper) (Zhou et al. 2011)) and Scophthalmids (flounder) (Irono et al. 2005)), but also in another fish species (for review, see Shi and Camus 2006). Hepcidins exhibit a dual role, being both an antimicrobial effector and an iron metabolism regulator.

Hepcidins are cationic, displaying a net charge of +2 at physiological pH, amphipathic and exhibit a β-hairpin structure constricted by three cystine residues. Lauth et al. have shown by NMR that hepcidins from hybrid striped bass presented almost the same structure as human hepcidins (Lauth et al. 2005). Indeed, the two AMP families have displayed an antiparallel β-sheet tightly held by a three inter-strand disulfide bond pattern. Both human and fish hepcidins exhibit an uncommon vicinal disulfide bridge (Lauth et al. 2005). To our knowledge, no structure-activity relationship studies of fish hepcidins have been carried out to date. On the other hand, they have been investigated in frog and human hepcidins. Human hepcidin has been shown to strongly bind DNA while Cys substitution by Leu in a synthetic analog resulted in a total loss of DNA binding abilities and antimicrobial activity (Hocquellet et al. 2012). Antimicrobial activity was also lost with reduced synthetic analogs in frog hepcidin whenever the cysteine signature was affected (Chanu et al. 2011).

Fish hepcidins have shown a narrower antimicrobial spectrum than pleurocidins, being only active against gram-negative bacteria and fungi. Moreover, they seem to be less potent than the latter with an average MIC estimated to be about 100 μg.ml^{-1} (Lauth et al. 2005). Furthermore, no haemolytic activity has been detected up to the MIC against hybrid sea bass erythrocytes even after four hours of incubation (Lauth et al. 2005). They have been shown to act synergistically with moronecidins (Lauth et al. 2005). While their antimicrobial activity appears less effective, their involvement in the fish defence system definitely emerges. Indeed, hepcidin gene expression is dramatically up-regulated under bacterial challenge with both firmicute and gracillicute sea bass pathogen within hours following bacterial infection (Shike et al. 2002, Lauth et al. 2005).

Sea bass hepcidin is produced as a large prepropeptide (85-residues long) including a 21-residue long signal peptide, a prodomain and the mature peptide of 21 residues (Shike et al. 2002). Despite the fact that sea bass hepcidin has been isolated from gills, major gene expression has been demonstrated in liver (Shike et al. 2002, Lauth et al. 2005). The authors hypothesized that hepcidin synthesized in liver may travel *via* humoral circulation to sites such as the gills to defend against invasion (Shike et al. 2002, Lauth et al. 2005). Unlike pleurocidins, hepcidins have not been detected in sea bass skin in spite of the same purification procedures (Lauth et al. 2002).

A unique and original fact about AMPs from eukaryotes is that hepcidins possess dual functions, being antimicrobial compounds actively involved in defence mechanisms as previously seen and iron-regulatory hormones (Rodrigues et al. 2006, Shi and Camus 2006). This peptide family constitutes a link between defence mechanisms and the iron metabolism (for review see Atanasiu et al. 2007).

Antimicrobial Peptide Fragments from Marine Sources

For a few years, AMPs resulting from *in vivo* proteolytic cleavage of inactive protein have stood out from classic AMPs, namely those synthesized as prepropeptide. It has been demonstrated that they were widely distributed in the living kingdom. Indeed, they have been isolated from organism ranging from invertebrates to superior mammals such as human beings. In this part, we have decided to describe only those arising from native functional protein excluding the cathelicidin family of AMPs (Shinnar et al. 2003), chromogranin fragments (Tota et al. 2007) and fragments resulting from neuropeptide-precursor maturation. For more information on the latter, the reader is referred to several reviews (Metz-Boutigue et al. 2003, Brogden et al. 2005, Salzet 2005). This part therefore focuses on *in vivo* enzymatic peptidic fragments exhibiting antimicrobial activity that we have chosen to call antimicrobial peptide fragments (AMPf) since they result from an inactive (at the antibiotic level) and functional protein involved in different roles.

As few post-peptides have been described, we have decided not to restrict this analysis to fish species but to extend it to aquatic species.

Post-peptides have been isolated from Crustaceans (shrimp (Destoumieux-Garzon et al. 2001) and crayfish (Lee et al. 2003)), Teleost fish (catfish (Park et al. 1998b), halibut (Birkemo et al. 2003), cod (Bergsson et al. 2005), salmon (Richards et al. 2001)) and Amphibians (toad) (Anderson and Beaven 2001). Other putative post-peptides have been predicted in shrimps (Patat et al. 2004), scallops (Li et al. 2007) and sea cucumbers (Hatakeyama et al. 2004). These post-peptides have been shown to derive from two main proteins, histones and hemocyanin.

AMPf from Histones

Histones are heat-resistant highly cationic proteins involved in chromatine structure. They are divided into two sub-classes according to their mobility. So, core histones composed of H2A, H2B, H3, and H4, are relatively immobile in comparison with linker histones called H1 and H5 (Parseghian and Luhrs 2006). We have distinguished AMPf resulting from proteolytic cleavage of core histones from those ensuing from linker histones.

AMPf from core histones, AMPf from core histones are presented in Table 5. The oldest one, a 39-residue long peptide named buforin was isolated from the stomach tissue of an Asian toad, *Bufo bufo gargarizans* (Park et al. 1996). Two years later, the same authors described a 19-mer AMPf called parasin I, in the epithelial mucosal layer of a Catfish, *Parasilurus asotus* (Park et al. 1998b). Hipposin, a 51-mer AMPf, was purified and characterized from skin mucus of halibut, *Hippoglossus hippoglossus* (Birkemo et al. 2003). The three AMPfs have been shown to derive from histone H2A. To date, in invertebrates, only putative AMPfs from histone H2A have been described in shrimp (Patat et al. 2004) and scallop (Li et al. 2007). They were shown to possess important sequence homologies with the former. Other antimicrobial fragments derived from histone H2A have been isolated especially from rainbow trout, *Oncorhyncus mykiss* (Fernandes et al. 2002). They will not be described in this review because they are proteins (MM~10 kDa).

Buforin, parasin and hipposin have demonstrated broad and potent antibacterial activities against both gram-positive and gram-negative bacteria (Park et al. 1996, Park et al. 1998b, Birkemo et al. 2003, Birkemo et al. 2004). Parasin was also shown to be antifungal but no haemolytic activity was detected against human erythrocyte up to 100µM (Park et al. 1998b). Using CD studies and structure prediction, parasin I was shown to adopt an amphiphilic β-strand structure while hipposin seemed to adopt an amphipathic and a partially α-helical organization (Park et al. 1998b, Birkemo et al. 2004). Despite various origins and lengths, the three H2A-derived AMPfs have displayed sequence homologies suggesting a common antimicrobial sequence (Table 5). Indeed, parasin and buforin display strong sequence homologies with respectively the N-terminal and the central part

Table 5 Antimicrobial peptide fragments isolated from aquatic species.

Name	Producer	Fragment	Sequence	Activity	References
Oncorhyncins					
Oncorhyncin-I	*O. mykiss*	Histone H1	SKGKKANKDVELARG	Gram+	(Fernandes et al. 2004)
Oncorhyncin II	*O. mykiss*	Histone H1	KAVAAKKSPKKAKKPATPKKAAKSPKKVKKPAAAAKKAAKSPKKATKAAKPKAAKPKAAKAKKAAPKKK	Gram+, Gram-	(Fernandes et al. 2004)
Oncorhyncin III	O. mykiss	Protein H6	PKRKSATKGDEPARRSARLSARPVPKPAAKPKKAAAPKKAVKGKKAAENGDAKAEAKVQAAGDGAG	Gram+,Gram-	(Fernandes et al. 2003)
SAMP H1	*Salmo salar*	Histone H1	AEVAPAPAAAAPAKAPKKKAAAKPKKAGPS	Gram+, Gram-	(Luders et al. 2005)
Buforin					
Buforin I	*Bufo gargarizans*	Histone H2A	AGRGKQGGKVRAKAKTRSSRAGLQFPVGRVHRLLRKGNY	Gram+, Gram-/ Fungus	(Park et al. 1996)
Buforin II	*Bufo gargarizans*	Histone H2A	TRSSRAGLQFPVGRVHRLLRK	Gram+, Gram-	
Parasin					
Parasin I	*Ps. asotus*	Histone H2A	KGRGKQGGKVRAKAKTRSS	Gram+, Gram-/ Fungus	(Park et al. 1998b)
Hipposin					
Hipposin	*H. hippoglossus*	Histone H2A	SGRGKTGGKARAKAKTRSSRAGLQFPVGRVHRLLRKGNYAHRVGAGAPVYL	Gram+, Gram-	(Birkemo et al. 2003)
Astacidins					
Astacidin 1	*Pf. leniusculus*	Hemocyanin	RKVQNQHGQVVKIFHH	Gram+, Gram-	(Lee et al. 2003)
Astacidin 2	*Pf. leniusculus*	Hemocyanin	RPRPNYRPRPIYRP	Gram+, Gram-	(Jiravanichpaisal et al. 2007)
PvHct	*P. vannamei*	Hemocyanin	FEDLPNFGHIQVKVFNHGEHIHH	Fungus	(Destoumieux-Garzon et al. 2001)
PsHct1	*P. stylirostris*	Hemocyanin	VTDGDADSAVPNLHENTEYNHYGSHGVYPDK	Fungus	(Destoumieux-Garzon et al. 2001)
PvHct2	*P. stylirostris*	Hemocyanin	LVVAVTDGDADSAVPNLHENTEYNHYGSHGVY	Fungus	(Destoumieux-Garzon et al. 2001)

Abbreviations *O., Ps., H., P.* and *Pf.* reffer respectively to the genera *Oncorhynchus, Parasilurus, Hippoglossus, Penaeus and Pacifastacus.*

of hipposin. Structure-activity relationship studies using synthetic peptides were not able to identify a minimal antimicrobial sequence. Both N-terminal (hip 1–19) and central (hip 16–36) fragments of hipposin have demonstrated antibacterial activities (Birkemo et al. 2004). Moreover, the main N-terminal part (hip 1–39) has displayed similar antibacterial activities on the 20 target bacteria (Birkemo et al. 2004).

These AMPfs' mechanism of action was investigated through buforin/lipid bilayer interactions and it was shown that buforin interacts with and crosses the plasmatic membrane of target cells to strongly bind to nucleic acids without dramatic membrane permeabilization (Park et al. 1998a, Kobayashi et al. 2004). In the light of sequence homologies and synthetic fragment activities detailed previously, it appears risky to extrapolate the mechanism of action of buforin to other H2A-derived AMPfs.

Little information is available concerning the generation of H2A-derived AMPfs. Parasin was isolated from catfish presenting epidermal injury while no parasin was produced by healthy catfish (Park et al. 1998b), suggesting that injuries or infections induce histone H2A proteolysis. Using immunohistochemical analyses and cDNA cloning, Cho et al. have revealed that most histone H2A, the parasin precursor, accumulates in the cytoplasm (Cho et al. 2002a), then is processed by cathepsin D itself activated by a matrix metalloproteinase (Cho et al. 2002b). On the other hand, relations between epidermal injury and precedent protease still remain to be elucidated.

AMPf from linker histones, Two AMPfs arising from H1-histone fragments have been reported from two *Oncorhynchus* sp., *O. kisutch* (atlantic salmon) and *O. mykiss* (rainbow trout). These two AMPfs have been shown to correspond respectively to the N-terminus and C-terminus of H1 histone (Fernandes et al. 2004, Luders et al. 2005). The former, named SAMP H1, is a 30-residue long acetylated peptide. Its amino acid composition is original since it contains about 40% of alanine, 23% of lysine and 20% of proline residues (Table 5). This latter residue has been shown to play a crucial role for SAMP H1 antibacterial activity. Indeed, it has appeared that synthetic peptide needs to be activated by a peptidylproline isomerase to gain antibacterial activity (Luders et al. 2005). For the first time, the role of proline isomers in AMP activity has been highlighted.

Oncorhyncin II is a 69-residue long antibacterial peptide arising from the C-terminal end of histone H1 (Fernandes et al. 2004). Both AMPfs from fish histone have been shown to be active against both Gram-negative and Gram-positive bacteria (Fernandes et al. 2004, Luders et al. 2005). In both cases, no fungal strains were assayed. For oncorhyncin II, no haemolytic activity was detected against trout haemocytes. To date, no elements are available concerning the regulation of SAMP H1 and Oncorhyncin II generation.

AMPfs from Hemocyanins

Hemocyanins are found in hemolymph of molluscs and arthropods. These oxygen transporters are giant extra cellular copper-containing proteins. Their structure is very different between the two phyla. Arthropod hemocyanins are composed of subunits of 72 kDa. The subunits are organized to form multiple hexamers (1 hexamer up to 8 hexamers) according to the species. Mollusc hemocyanins are different since 7 or 8 functional units of 50 kDa form a subunit of hemocyanin. Ten subunits have been shown to associate forming a decameric or didecameric hollow cylinder depending on the species. For review, the reader is referred to the work of Decker and Jaenicke (2004) and van Holde et al. (2001).

AMPs from proteolytic cleavage of hemocyanin have been described (Table 5). Both of the AMPs identified were obtained from Crustaceans, one from the freshwater crayfish *Pacifastacus leniusculus* (Lee et al. 2003) and the other from shrimps *Peneaus vannamei* and *P. stylirostris* (Destoumieux-Garzon et al. 2001).

Destoumieux-Garzon et al. have characterized a 23-mer AMP named PvHct from *P. vannamei* while PsHCt1 and PvHCt2, respectively 31- and 32-residue long AMPfs, were purified from *P. stylirostris* (Destoumieux-Garzon et al. 2001). Sequence comparisons have demonstrated that all these peptides arose from the C-terminal part of hemocyanin. Moreover, they were original in so far as (i) they exhibited an exclusive anti-fungal activity and (ii) they displayed an anionic net charge at physiological pH. Immune challenge of shrimps resulted in the detection of a higher concentration of PvHCt in shrimp plasma (Destoumieux-Garzon et al. 2001).

Similar results were obtained from the freshwater crayfish where astacin, a 16-residue long AMP, was detected and characterized. Astacin has been shown to derive from the C-terminal part of hemocyanin (Lee et al. 2003). Astacin has exhibited a broad spectrum of antibacterial activity. Truncated synthetic peptides have clearly demonstrated the crucial role of N-terminal residues in antibacterial activity. CD studies have shown that astacin adopts a β-strand structure (Lee et al. 2003). Like for AMPfs from shrimps, immune challenges in the form of lipopolysaccharide or glucan injection resulted in an increase of astacin production (Lee et al. 2003).

Hemocyanin cleavage yielding AMPfs in Crustaceans could therefore not only result from extraction procedures but from biological processes involved in innate immunity. It is hypothesized that Crustaceans under assault took advantage of an abundant protein to rapidly generate AMPfs from their C-terminal part. Moreover, hemocyanins are involved in innate immunity *via* other mechanisms. Indeed, it has been demonstrated that hemocyanin, an O_2 transporter, was converted to phenoloxydase, an enzyme involved in defence mechanisms (Nagai et al. 2001, Adachi et al. 2003). It appears that the giant protein may play various roles in fighting infection by generating AMPfs and / or phenoloxidase activity. Moreover, hemocyanin has been shown to demonstrate antiviral activity (Zhang et al. 2004a).

Bacteriocins: AMPs Produced by Bacteria

Aquatic animals are in permanent contact with their surrounding microflora. Bacteria can maintain and grow independently from a host, so they can occur in high densities around animals (Verschuere et al. 2000). Furthermore, these microorganisms can develop inside animals as well, entering the host during feeding, and after ingestion they influence the indigenous intestinal flora. In fish, bacterial density was measured in gill tissues at 10^6 bacteria.g^{-1}, on skin between 10^2 and 10^4 bacteria.cm^{-2}, and in the digestive tract up to 10^{13} bacteria.g^{-1} (Austin 2006). Depending on their ability to colonize the host, fish-associated bacteria are divided into two categories: indigenous (or autochtonous) bacteria can multiply inside the host, especially in the gastrointestinal tract, whereas exogenous (or allochtonous) bacteria cannot. Classical microbiology culture techniques, such as dilution, spread plating or swabbing, are commonly used to determine the fish-associated microflora. Consequently, the indigenous origin of the isolated bacteria can be subject to debate (Austin 2006).

Like in mammalians, indigenous flora can exert beneficial effects on their host. Fish-associated bacteria are recognized to play different roles: (i) assist fish in locomotion by changing their surface properties, particularly for fast-swimming fish (Cahill 1990), (ii) protect the host from pathogenic strains directly by producing a high number of biologically active compounds like antimicrobial compounds, enzymes or siderophores, (iii) contribute to nutrition, especially for indigenous intestinal microbiota and (iv) enhance the immune system of their host (Chiller et al. 2001, Silva et al. 2005, Austin 2006, Nayak 2010). The ability of fish to tolerate or select bacteria with antibacterial activity leads us to look for AMPs produced by these microorganisms. Unfortunately, most of the studies aim at selecting potential probiotics for aquaculture farming. Consequently, only poor structural information is generally available on AMPs produced by fish-associated bacteria. In this section we focus on antimicrobial compounds for which the proteinaceous nature was established or suspected. AMPs produced by bacteria are usually named bacteriocins which can be defined as follows: "ribosomally synthesized proteinaceous compounds, lethal to bacteria closely related to the producing bacteria, the latter being protected by an immunity phenomenon" (Desriac et al. 2010). Bacteriocins are divided into two groups contingent on their size: protein bacteriocins and peptide bacteriocins.

AMP-like from indigenous gut bacteria

Investigations on gut microflora from freshwater or marine fish led to the isolation of bacteria from various genera, such as *Acitenobacter, Enterobacter, Escherichia, Carnobacterium, Klebsellia, Pseudomonas, Aeromonas,* or *Vibrio* (Austin 2006, Nayak 2010). Lactic Acid Bacteria (LAB) are a well-known kind of gut-associated bacteria that could bring a significant advantage to their hosts, particularly as concerns resistance to pathogens—for review see Ringø and Gatesoupe 1998. Furunculosis is one of those common diseases caused by microbial agents such as *Aeromonas salmonicida*. Balcázar et al. (2007) evidenced an enhancement of the survival rate of rainbow trouts (*Oncorhynchus mykiss*) exposed to this pathogen by a simple oral administration of LAB. Hence, they underlined how important LAB could be for the health of their hosts. Several studies found that LAB in the fish gut displayed antibacterial activity (Spanggaard et al. 2001, Silva et al. 2005, Balcázar et al. 2007, Kim and Austin 2008, Nayak 2010, Sequeiros et al. 2010, Pérez-Sánchez et al. 2011, Satish Kumar et al. 2011). However, only few of them revealed the nature of the active compounds.

One of the first reports on fish-associated LAB was published by Stoffels et al. in 1992. From a *Carnobacterium* sp. culture supernatant, the authors used ammonium sulfate precipitation and several chromatographic steps to isolate an antibacterial compound termed carnocin UI49. This AMP is composed of at least 35 amino acids (not including the number of cysteins and possible residues of tryptophan) for a molecular weight of 4,635 Da. The mechanism of action of carnocin UI49 was investigated using the sensitive strain *Lactococcus lactis* subsp. *Lactis* and mutants (Stoffels et al. 1994). The bactericidal properties of carnocin are due to their interaction with the cytoplasmic membrane, probably through a receptor specific for nisin-producing strains.

In 1995, two other carnobacteria (*C. piscicola* V1 and *C. divergens* V41) were isolated from fish for their activity against the human pathogen *Listeria monocytogenese* (Pilet et al. 1995). The purification of AMPs produced by these two LABs was performed using a two-step protocol, i.e., supernatant ammonium sulfate precipitation followed by reverse phase chromatography. The amino acid compositions of AMPs were determined using Edman degradation (Bhugaloo-Vial et al. 1996, Métivier et al. 1998). Analyses on *C. piscicola* V1 supernatant revealed two bacteriocins named piscicocins V1a and V1b with molecular masses of 4416 and 4526 Da, respectively. Piscicocin V1a and V1b belong to class IIa bacteriocins but exhibited significant differences in their inhibitory activities against gram positive bacteria. The specific activity of V1a was around 100 times higher than that of V1b. Piscicocins have a bactericidal activity by insertion in the lipid bilayers. Investigations on *Carnobacterium divergens* V41 demonstrated the production by this strain of a new class IIa bacteriocin composed of 43 amino acids for a molecular mass of 4513 Da (for review on Divercin V41, see Rihakova et al. 2009).

More recently, two studies were led to isolate anti-*Lactococcus garviae* strains from the intestinal tract of fish. The indigenous flora from the rainbow trout *Oncorhynchus mykiss* (Pérez-Sánchez et al. 2011) and from the patagonian fish *Odontesthes platensis* (Sequeiros et al. 2010) was investigated. Pérez-Sánchez et al. (2011) screened bacteria on non-selective medium TSB (Tryptone Soy Broth) and selective MRS (De Man Rogosa and Sharpe) agar, while Sequeiros et al. (2010) focused only on LAB using MRS broth. Pérez-Sánchez et al. (2011) demonstrated that over 3% of the isolated bacteria exerted an inhibitory activity against the fish pathogen *L. garviae*. When only LAB were isolated, this percent increased about 2.5 fold (7.5%). After isolation and screening of the active bacteria, both studies adopted the same approach. First, they chose to neutralize the supernatant in order to exclude pH interferences while determining its activity. In the second stage, they treated the supernatant with proteinase K so as to determine the proteinaceous nature of active compounds. Further experiments on the active molecule by Sequeiros et al. (2010) revealed a high stability of the bacteriocin, especially regarding temperature (resistant to 121°C for 15 min) and pH range (from 3 to 11). Pérez-Sánchez et al. (2011) highlighted the production of AMPs by three different LAB isolated from rainbow trout (*Lactobacillus mesenteroides, Lactococcus lactis lactis* and *Lactobacillus plantarum*), whereas Sequeiros et al. (2010) evidenced an AMP production for only one strain, *Lactococcus lactis* TW34, isolated from Patagonian fish. The antimicrobial spectrum of this strain was investigated further and the authors established that it was only active against gram positive bacteria (*Carnobacterium, Listeria, Lactococcus* and *Streptococcus*) supporting the bacteriocin nature of active compounds.

From the gut of grey mullet fish (*Mugil cephalus*), Satish Kumar et al. (2011) isolated *Enterococcus faecium*. Using ultrafiltration, precipitation and chromatographic techniques, they subsequently purified a peptide named enterocin MC13 which has a molecular mass of 2.148 kDa (MALDI-TOF analysis). This peptide exhibited an inhibitory activity against 12 of the 14 indicator strains tested. Enterocin MC13 had an antibacterial activity against both gram negative bacteria (*Vibrio* and *Aeromonas* species) and gram positive bacteria (*Listeria monocytogenese* and *Lactobacillus* species).

AMP-like from gill- or skin-associated bacteria

Skin-associated bacteria include Acinetobacter, Aeromonas, Flavobacterium, Micrcoccus and Vibrio (Spanggaard et al. 2001, Austin 2006). Proteobacteria are generally found in high proportions on skin microbiota (Boutin et al. 2012). For example, it was shown that this bacterial class is the second group isolated from the skin of brook charr (Salvelinus fontinalis) whereas it was the first from rainbow trout (Onchorhynchus mykiss) with about 71% of strains isolated from skin and gills displayed in the gamma proteobacteria class (Spanggaard et al. 2001). The predominant genera appear to be Pseudomonas, Aeromonas, Moraxella and Vibrio (Cahill 1990). Actually, excluding Escherichia bacteriocins, the number of reports on AMPs produced by gamma proteobacteria is significantly lower than those from gram positive bacteria. As an example, in Bactibase (see Table 1) 166 bacteriocins from firmicutes are described, whereas only 5 are listed for gammaproteobacteria (17 if Escherichia bacteriocins are included). Among them, only the genera Serratia and Pseudomonas are known to produce bacteriocins (Serratin P and Pyocins bacteriocins, respectively).

Concerning the genus *Vibrio*, Zai et al. (2009) focused on *Arius thalassimus* in order to discover indigenous *Vibrio* spp. producing Bacteriocin-like Inhibitory Substances (BLIS). After bacteria selection, their antibacterial activity was tested by two distinct methods: stab-overlay and agar-well diffusion. The latter brought to light 12 strains that produced BLIS whereas the first allowed the isolation of 20 strains. These results put the emphasis on techniques used during bacteria screening. The antibacterial spectrum of one strain isolated (AVP10) was established according to the growth medium and it turned out to be active only against four strains of *Vibrio* for the LB medium. A supplementary sensitive strain (*E. coli*) was found when the production medium was a citrate phosphate nutrient agar medium. This point highlights the impact of culture medium for the discovery of new AMPs produced by bacteria.

It is surprising that since the first report (Farkas-Himsley and Seyfried 1962) on *Vibrio* species producing BLIS in the beginning of the 1960s, little structural information has become available for vibriocin substances (Chakrabarty et al. 1970, McCall and Sizemore 1979, Shehane and Sizemore 2002, Carraturo et al. 2006). The most comprehensive report on vibriocins was proposed by Prasad et al. (2005). In this study, the secretion of a 32 kDa protein by *V. harveyi* VIB571 was described (molecular weight determined by SDS-PAGE). N-terminal sequencing established the first 14 amino acids of the BLIS protein: Asp-Glu-Tyr-Ile-Ser-X-Asn-Lys-X-Ser-Ser-Ala-Asp-Ile. The inhibitory spectrum of strain VIB571 was shown to be closely related to *Vibrio* strains such as *V. harveyi* strains, *V. fischeri* or *V. gazogenes*. No inhibitory activity was detected against methicillin-resistant *Staphylococcus aureus* strain and vancomycin-resistant enterococci.

Strain *Vibrio* sp. NM 10 was isolated from spotnape ponyfish (*Leiognathus nuchalis*) collected in the coastal regions of Enoshima Island, Kanagawa, Japan. It exhibited high activity against *Pasteurella piscicida* K-III, but was also able to inhibit *E. coli* IAM 1264, *V. vulnificus* RIMD 2219009 and *Enterococcus seriolicida* YT-3. The antibacterial substance produced by *Vibrio* sp. NM 10 is a proteinaceous heat-labile substance with a molecular mass of less than 5 kDa (determined by ultrafiltration). These facts strongly suggest that the antibacterial substance is either a bacteriocin or a bacteriocin-like substance (Sugita et al. 1997).

Need for activity research and structural data in fish-associated bacteria

In general, investigating the associated microflora from freshwater or marine fish leads to the isolation of different bacterial genera like *Acinetobacter, Enterobacter, Escherichia, Carnobacterium, Klebsellia, Pseudomonas, Aeromonas,* or *Vibrio* (Nayak 2010, Austin 2006). Among these genera, several species are known to produce bacteriocins or bacteriocin-like substances.

The most studied are certainly the bacteriocins produced by *Escherichia* species with two classes of microcins described (Jack and Jung 2000, Duquesne et al. 2007). Other bacteriocins are described in this bacterial family but they seem to have higher molecular masses, between 40 kDa and 80 kDa in particular for colicin-like bacteriocins (see Cascales et al. 2007 for review).

Enterococcus sp. secreted different bacteriocins generally called enterocins. At least 8 enterocins were described for an *E. faecium* species alone with masses comprising between 2.148 (Satish Kumar et al. 2011) and 6.39 kDa (Izquierdo et al. 2008). A high number of enterocins were characterized and turned out to have regular structural or genetic characteristics that allowed them to be classified using the current LAB bacteriocin classification system (Klaenhammer 1993). The structural diversity of some enterocins is such that Franz et al. (2007) proposed a simplified classification for enterocins. This classification includes three classes where class II is subdivided into three subclasses (Franz et al. 2007).

Pyocins are bactericocins produced by the majority of *Pseudomonas aeruginosa* strains. Three types are described: (i) R-type pyocins (ii) F-type pyocins and (iii) S-type pyocins (for review see Michel-Briand and Baysse 2002).

Most of the fish-associated strains belong to these genera that are known to produce AMPs. Also, these bacteria have been described by many authors, the search for antimicrobial activity has not been undertaken extensively, and when it has, structural data about active compounds are often missing. Nevertheless, we can reasonably assume that many of these bacteria might play a significant role regarding their host, producing AMPs or other kinds of antimicrobial substances. Analyses of antimicrobial properties of fish-associated bacteria and structural determination of the active compounds are a promising challenge for research in the next years, as today this kind of data is particularly scarce.

CONCLUSION

Most fish species are known to produce AMPs. These fish-produced AMPs can be classified into four subgroups based on their structural features or their origins:

1. **Linear helical AMPs:** These AMPs have mainly been found in skin or gills of fish. They exhibit potent antimicrobial activity and a broad spectrum of action, being active against gram-positive and gram-negative bacteria. Fungicidal and haemolytic activities have also been described for some of them. They display common structural features with other vertebrate AMPs. In addition to antimicrobial activity, these peptides have been shown to be involved in immune function modulation suggesting a more important role than a potent chemical weapon.
2. **Cyclic cystein-rich AMPs:** They share a cyclic structure tightly held by several disulfide bridges. The cysteine signature has been shown to be crucial for antimicrobial activity. Based on constrained structure, we can reasonably assume that they may display great stability against physicochemical denaturation and maybe enzymatic cleavage. Their antimicrobial activity seems to be both less potent and more specific than the former, although some of them show a potent antifungal activity. It is risky to compare antimicrobial activity performed in different conditions with distinct strains. Like helical AMPs, cyclic cystein-rich AMPs also exhibit a dual role, as they are also involved in iron metabolism.
3. **AMP fragments of abundant protein:** These peptides resulted from structural or functional protein hydrolysis. In response to microbial assault, aquatic species may use protein in large amounts to generate AMPs. These AMPs are believed to complement defence effectors. They exhibit overall common structural features with gene-encoded AMPs being cationic 20 to 60-residue long peptides.
4. **Bacteriocins:** On top of occupying the ecological niche, some bacteria have been shown to produce BLIS. Only a few have been characterized from fish-isolated producing strains. BLIS-producing strains have been isolated from environmentaly-exposed tissues, e.g., skin, gills or the gastro-intestinal tract. Such a BLIS production can be considered as a probiotic trait since it is likely to provide a first natural defence barrier against pathogenic micro-organisms.

All these sub-groups have been described in the same fish species, for instance *Oncorhynchus mykiss*, illustrating the various levels of fish defense.

Faced with these data, different and complementary research avenues should be explored to reduce the impact of infection on intensive animal breeding and as a consequence the use of antibiotic treatments:

Multiple role of antimicrobial peptides: Mammal AMPs have been renamed host defence peptides due to their immunomodulatory activity. Indeed they have been shown to influence cellular functions by increasing or reducing gene transcription, cytokine production, chemotaxis and apoptosis (Wiesner and Vilcinskas 2010). Several works have shown that fish AMPs play a similar role (for review see Rajanbabu and Chen 2011). In terms of animal breeding, genetic selection of fish based on AMP production may result in animals that are more resistant to infections and stress.

Probiotic selection based on BLIS-production from healthy wild fish: We have already presented a strategy aimed at selecting potential probiotics based on BLIS-producing abilities from among indigenous bacteria associated with marine organisms (Desriac et al. 2010). A probiotic collection exhibiting antimicrobial activity against pathogenic stains may arise from such an approach.

Another point which should be underlined is Fish AMP production is mainly located in environmentally exposed tissues such as gills, skin and the gastrointestinal tract. All of these organs constitute waste in the fishery industry. This biomass is not exploited except for fish meal. Taking into account the exceptional potential of AMPs as active substances, an innovative technological network should be thought up to convert this biomass.

REFERENCES

Aboudy, Y., E. Mendelson, I. Shalit, R. Bessalle and M. Fridkin. 1994. Activity of two synthetic amphiphilic peptides and magainin-2 against Herpes simplex virus types 1 and 2. Int. J. Pept. Prot. Res. 43(6): 573–582.

Adachi, K., T. Hirata, T. Nishioka and M. Sakaguchi. 2003. Hemocyte Components in Crustaceans Convert Hemocyanin into a Phenoloxidase-like Enzyme. Comp. Biochem. Physiol. 134: 135–141.

Amiche, M., A.A. Seon, T.N. Pierre and P. Nicolas. 1999. The Dermaseptin Precursors: a Protein Family with a Common Preproregion and a Variable C-terminal Antimicrobial Domain. FEBS J. 456: 352–356.

Anderson, R.S. and A.E. Beaven. 2001. Antibacterial Activties of Oyster (Crassostrea virginica) and Mussel (Mytilus edulis and Geukensia demissa) Plasma. Aquat. Living Resour. 14: 343–349.

Andreu, D. and L. Rivas. 1998. Antimicrobial Peptides : an Overview. Biopolymers. 47: 415–433.

Atanasiu, V., B. Manolescu and I. Stoian. 2007. Hepcidin-Central Regulator of Iron Metabolism. Eur. J. Haematol. 78(1): 1–10.

Austin, B. 2006. The Bacterial Microflora of Fish, Revised. Scientific World Journal. 6: 931–945.

Baghian, A., J. Jaynes, F. Enright and K. G. Kousoulas. 1997. An Amphipathic α-Helical Synthetic Peptide Analogue of Melittin Inhibits Herpes Simplex Virus-1 (HSV-1)-Induced Cell Fusion and Virus Spread. Peptides. 18(2): 177–183.

Balcázar, J.L., D. Vendrell, I. de Blas, I. Ruiz-Zarzuela, O. Gironés and J.L. Múzquiz. 2007. In vitro Competitive Adhesion and Production of Antagonistic Compounds by Lactic Acid Bacteria Against Fish Pathogens. Vet. Microbiol. 122(3-4): 373–380.

Bao, B., E. Peatman, P. Li, C. He and Z. Liu. 2005. Catfish Hepcidin Gene is Expressed in a Wide Range of Tissues and Exhibits Tissue-specific Upregulation After Bacterial Infection. Dev. Comp. Immunol. 29(11): 939–950.

Bergsson, G., B. Agerberth, H. Jornvall and G.H. Gudmundsson. 2005. Isolation and Identification of Antimicrobial Components from the Epidermal Mucus of Atlantic Cod (Gadus morhua). FEBS J. 272(19): 4960–4969.

Bhatnagar, I. and S.-K. Kim. 2010. Immense Essence of Excellence: Marine Microbial Bioactive Compounds. Mar. Drugs. 8(10): 2673–2701.

Bhugaloo-Vial, P., X. Dousset, A. Metivier, O. Sorokine, P. Anglade, P. Boyaval and D. Marion. 1996. Purification and Amino Acid Sequences of Piscicocins V1a and V1b, Two Class IIa Bacteriocins Secreted by Carnobacterium Piscicola V1 That Display Significantly Different Levels of Specific Inhibitory Activity. Appl. Environ. Microbiol. 62(12): 4410–4416.

Birkemo, G.A., T. Lüders, Ø. Andersen, I.F. Nes and J. Nissen-Meyer. 2003. Hipposin, a Histone-derived Antimicrobial Peptide in Atlantic Halibut (Hippoglossus hippoglossus L.). Biochim. Biophys. Acta. 1646(1-2): 207–215.

Birkemo, G.A., D. Mantzilas, T. Lüders, I.F. Nes and J. Nissen-Meyer. 2004. Identification and Structural Analysis of the Antimicrobial Domain in Hipposin, a 51-mer Antimicrobial Peptide Isolated from Atlantic Halibut. Biochim. Biophys. Acta. 1699(1-2): 221–227.

Blunt, J.W., B.R. Copp, M.H.G. Munro, P.T. Northcote and M.R. Prinsep. 2004. Marine Natural Products. Nat. Prod. Rep. 21: 1–49.

Boutin, S., L. Bernatchez, C. Audet and N. Derôme. 2012. Antagonistic Effect of Indigenous Skin Bacteria of Brook Charr (Salvelinus fontinalis) Against Flavobacterium columnare and F. psychrophilum. Vet. Microbiol. 155(2-4): 355–361.

Brogden, K.A. 2005. Antimicrobial Peptides: Pore Formers or Metabolic Inhibitors in Bacteria? Nat. Rev. Micro. 3(3): 238–250.

Brogden, K.A., J.M. Guthmiller, M. Salzet and M. Zasloff. 2005. The Nervous System and Innate Immunity: The Neuropeptide Connection. Nat. Immunol. 6: 558–564.

Browne, M.J., C.Y. Feng, V. Booth and M.L. Rise. 2011. Characterization and Expression Studies of Gaduscidin-1 and Gaduscidin-2; Paralogous Antimicrobial Peptide-like Transcripts from Atlantic Cod (Gadus morhua). Dev. Comp. Immunol. 35(3): 399–408.

Cahill, M.M. 1990. Bacterial Flora of Fishes: A Review. Microb. Ecol. 19(1): 21–41.

Campagna, S., N. Saint, G. Molle and A. Aumelas. 2007. Structure and Mechanism of Action of the Antimicrobial Peptide Piscidin. Biochemistry. 46(7): 1771–1778.

Carraturo, A., K. Raieta, D. Ottaviani and G.L. Russo. 2006. Inhibition of Vibrio parahaemolyticus by a Bacteriocin-like Inhibitory Substance (BLIS) Produced by Vibrio mediterranei 1. J. Appl. Microbiol. 101(1): 234–241.

Cascales, E., S.K. Buchanan, D. Duché, C. Kleanthous, R. Lloubès, K. Postle, M. Riley, S. Slatin and D. Cavard. 2007. Colicin Biology. Microbiol. Mol. Biol. Rev. 71(1): 158–229.

Chakrabarty, A.N., S. Adhya, J. Basu and S.G. Dastidar. 1970. Bacteriocin Typing of Vibrio cholerae. Infect. Immun. 1(3): 293 –299.

Chanu, K.V., A. Kumar and S. Kumar. 2011. Structure-activity Relationship of Buffalo Antibacterial Hepcidin Analogs. Indian J. Biochem. Biophys. 48: 325–330.

Charpentier, S., M. Amiche, J. Mester, V. Vouille, J.-P. Le Caer, P. Nicolas and A. Delfour. 1998. Structure, Synthesis and Molecular Cloning of Dermaseptins B, a Family of Skin Peptide Antibiotics. J. Biol. Chem. 273(24): 14690–14697.
Chekmenev, E.Y., B.S. Vollmar, K.T. Forseth, M.N. Manion, S.M. Jones, T.J. Wagner, R.M. Endicott, B.P. Kyriss, L.M. Homem, M. Pate, J. He, J. Raines, P.L. Gor'kov, W.W. Brey, D.J. Mitchell, A.J. Auman, M.J. Ellard-Ivey, J. Blazyk and M. Cotten. 2006. Investigating Molecular Recognition and Biological Function at Interfaces Using Piscidins, Antimicrobial Peptides from Fish. Biochim. Biophys. Acta. 1758(9): 1359.
Chiller, K., B.A. Selkin and G.J. Murakawa. 2001. Skin Microflora and Bacterial Infections of the Skin. The J. Invest. Dermatol. Symp. Proc. 6(3): 170–174.
Chinchar, V.G., L. Bryan, U. Silphadaung, E. Noga, D. Wade and L. Rollins-Smith. 2004. Inactivation of Viruses Infecting Ectothermic Animals by Amphibian and Piscine Antimicrobial Peptides. Virol. 323(2): 268–275.
Cho, J. and D.G. Lee. 2011. Oxidative Stress by Antimicrobial Peptide Pleurocidin Triggers Apoptosis in Candida albicans. Biochimie. 93(10): 1873–1879.
Cho, J.H., I.Y. Park, H.S. Kim, W.T. Lee, M.S. Kim and S.S. Kim. 2002a. Cathepsin D Produces Antimicrobial Peptide Parasin I from Histone H2A in the Skin Mucosa of Fish. FASEB J. 16(3): 429–431.
Cho, J.H., I.Y. Park, M.S. Kim and S.C. Kim. 2002b. Matrix Metalloproteinase 2 Is Involved in the Regulation of the Antimicrobial Peptide Parasin I Production in Catfish Skin Mucosa. FEBS Lett. 531(3): 459–463.
Cole, A.M., R.O. Darouiche, D. Legarda, N. Connel and G. Diamond. 2000. Characterization of a Fish Antimicrobial Peptide: Gene Expression, Subcellular Localization and Spectrum of Activity. Antimicrob. Agents Chemother. 44: 2039–2045.
Cole, A.M., P. Weis and G. Diamond. 1997. Isolation and Characterization of Pleurocidin, an Antimicrobial Peptide in the Skin Secretions of Winter Flounder. J. Biol. Chem. 272(18): 12008–12013.
Daher, K.A., M.E. Selsted and R.I. Lehrer. 1986. Direct Inactivation of Viruses by Human Granulocyte Defensins. J. Virol. 60(3): 1068–1074.
Decker, H. and E. Jaenicke. 2004. Recent Findings on Phenoloxydase and Antimicrobial Activity of Hemocyanins. Dev. Comp. Immunol. 28: 673–687.
Dennison, S.R., J. Howe, L.H.G. Morton, K. Brandenburg, F. Harris and D.A. Phoenix. 2006. Interactions of an Anionic Antimicrobial Peptide with Staphylococcus aureus Membranes. Biochem. Biophys. Res. Commun. 347(4): 1006–1010.
Desriac, F., D. Defer, N. Bourgougnon, B. Brillet, P. Le Chevalier and Y. Fleury. 2010. Bacteriocin as Weapons in the Marine Animal-Associated Bacteria Warfare: Inventory and Potential Applications as an Aquaculture Probiotic. Mar. Drugs. 8(4): 1153–1177.
Destoumieux-Garzon, D., D. Saulnier, J. Garnier, C. Jouffrey, P. Bulet and E. Bachere. 2001. Crustacean Immunity. Antifungal Peptides are Generated from the C Terminus of Shrimp Hemocyanin in Response to Microbial Challenge. J. Biol. Chem. 276(50): 47070–47077.
Devlieghere, F., L. Vermeiren and J. Debevere. 2004. New Preservation Technologies: Possibilities and Limitations. Int. Dairy J. 14: 273–285.
Dobson, A., P. Cotter, R. Ross and C. Hill. 2012. Bacteriocin Production: a Probiotic Trait? Appl. Environ. Microbiol. 78: 1–6.
Donia, M. and M.T. Hamann. 2003. Marine Natural Products and Their Potential Applications as Anti-infective Agents. Lancet Infect. Dis. 3(6): 338.
Douglas, S.E., J.W. Gallant, Z. Gong and C. Hew. 2001. Cloning and Developmental Expression of a Family of Pleurocidin-like Antimicrobial Peptides from Winter Flounder, Pleuronectes americanus (Walbaum). Dev. Comp. Immunol. 25(2): 137–147.
Douglas, S.E., J.W. Gallant, R.S. Liebscher, A. Dacanay and S.C.M. Tsoi. 2003a. Identification and Expression Analysis of Hepcidin-like Antimicrobial Peptides in Bony Fish. Dev. Comp. Immunol. 27(6–7): 589–601.
Douglas, S.E., A. Patrzykat, J. Pytyck and J.W. Gallant. 2003b. Identification, Structure and Differential Expression of Novel Pleurocidins Clustered on the Genome of the Winter

Flounder, Pseudopleuronectes americanus (Walbaum). Eur. J. Biochem. 270(18): 3720–3730.
Duquesne, S., D. Destoumieux-Garzón, J. Peduzzi and S. Rebuffat. 2007. Microcins, Gene-encoded Antibacterial Peptides from Enterobacteria. Nat. Prod. Rep. 24(4): 708–734.
Epand, R.M. and H.J. Vogel. 1999. Diversity of Antimicrobial Peptides and Their Mechanisms of Action. Biochim. Biophys. Acta. 1462: 11–28.
Farkas-Himsley, H. and P.L. Seyfried. 1962. Lethal Biosynthesis of a New Antibacterial Principle: Vibriocin. Nature. 193: 1193–1194.
Fernandes, J.M.O., G. Molle, G.D. Kemp and V.J. Smith. 2004. Isolation and Characterisation of Oncorhyncin II, a Histone H1-derived Antimicrobial Peptide from Skin Secretions of Rainbow Trout, Oncorhynchus mykiss. Dev. Comp. Immunol. 28(2): 127–138.
Fernandes, J.M.O., N. Saint, G.D. Kemp and V.J. Smith. 2003. Oncorhyncin III: a Potent Antimicrobial Peptide Derived from the Non-histone Chromosomal Protein H6 of Rainbow Trout, Oncorhynchus mykiss. Biochem. J. 373(2): 621–628.
Fernandes, J.M.O., G.D. Kemp, G. Molle and V.J. Smitth. 2002. Anti-microbial Properties of Histone H2A from Skin Secretions of Rainbow Trout, Oncorhynchus mykiss. Biochem. J. 368: 611–620.
Fischer, F.G. and W.P. Neumann. 1961. The Venom of the Honeybee. III. On the Chemical Knowledge of the Principle Active Constituent (melittin). Biochem Z. 335: 51–61.
Franz, C.M.A.P, M.J. van Belkum, W.H. Holzapfel, H. Abriouel and A. Gálvez. 2007. Diversity of Enterococcal Bacteriocins and Their Grouping in a New Classification Scheme. FEMS Microbiol. Rev. 31(3): 293–310.
García-Olmedo, F., A. Molina, J.M. Alamillo and P. Rodríguez-Palenzuéla. 1998. Plant Defense Peptides. Biopolymers. 47(6): 479–491.
Gross, E. and J.L. Morell. 1971. The Structure of Nisin. J. Am. Chem. Soc. 8: 4634–4635.
Hancock, R.E.W. and D.S. Chapple. 1999. Peptide Antibiotics. Antimicrob. Agents Chemother. 43(6): 1317–1323.
Hancock, R.E.W. and A. Patrzykat. 2002. Clinical Development of Cationic Antimicrobial Peptides: From Natural to Novel Antibiotics. Curr. Drug Targets Infect. Disord. 2: 79–83.
Hancock, R.E.W. and H.-S. Sahl. 2006. Antimicrobial and Host-defense Peptides as New Anti-infective Therapeutic Strategies. Nat. Biotech. 24(12): 1551–1557.
Hancock, R.E.W. and M.G. Scott. 2000. The Role of Antimicrobial Peptides in Animal Defenses. Proc. Nat. Acad. Sci. USA. 97: 8856–8861.
Hatakeyama, T, T. Suenaga, S. Eto, T. Niidome and H. Aoyagi. 2004. Antibacterial Activity of Peptides Derived from the C-Terminal Region of a Hemolytic Lectin, CEL-III, from the Marine Invertebrate Cucumaria Echinata. J. Biochem. (Tokyo). 135(1): 65–70.
Hilchie, A.L., C.D. Doucette, D.M. Pinto, A. Patrzykat, S.E. Douglas and D.W. Hoskin. 2011. Pleurocidin-family Cationic Antimicrobial Peptides Are Cytolytic for Breast Carcinoma Cells and Prevent Growth of Tumor Xenografts. Breast Canc. Res. 13: R102.
Hocquellet, A., C. le Senechal and B. Garbay. 2012. Importance of the Disulfide Bridges in the Antibacterial Activity of Human Hepcidin. Peptides. in press.
Iijima, N., N. Tanimoto, Y. Emoto, Y. Morita, K. Uematsu, T. Murakami and T. Nakai. 2003a. Purification and Characterization of Three Isoforms of Chrysophsin, a Novel Antimicrobial Peptide in the Gills of the Red Sea Bream, Chrysophrys major. Eur. J. Biochem. 270(4): 675–686.
Iijima, R., J. Kisugi and M. Yamazaki. 2003b. A Novel Antimicrobial Peptide from the Sea Hare Dolabella auricularia. Dev. Comp. Immunol. 27(4): 305–311.
Irono, I., J.Y. Hwang, Y. Ono, T. Ohira, R. Nozaki and T. Aoki. 2005. Two Different Types of Hepcidins from the Japanese Flounder, Paralichthys olivaceus. FEBS J. 272: 5257–5264.
Iwanaga, S. 2002. The Molecular Basis of Innate Immunity in the Horseshoe Crab. Curr. Op. Immunol. 14: 87–95.
Iwanaga, S. and B.L. Lee. 2005. Recent Advances in the Innate Immunity of Invertebrates Animals. J. Biochem. Mol. Biol. 38: 128–150.

Izquierdo, E., A. Bednarczyk, C. Schaeffer, Y. Cai, E. Marchioni, A. Van Dorsselaer and S. Ennahar. 2008. Production of Enterocins L50A, L50B, and IT, a New Enterocin, by Enterococcus faecium IT62, a Strain Isolated from Italian Ryegrass in Japan. Antimicrob. Agents Chemother. 52(6): 1917–1923.

Jack, R.W. and G. Jung. 2000. Lantibiotics and Microcins: Polypeptides with Unusual Chemical Diversity. Curr. Opin. Chem. Biol. 4(3): 310–317.

Jenssen, H., P. Hamill and R.E.W. Hancock. 2006. Peptide Antimicrobial Agents. Clin. Microbiol. Rev. 19(3): 491–511.

Jiravanichpaisal, P., S.Y. Lee, Y.-A. Kim, T. Andrén and I. Söderhäll. 2007. Antibacterial Peptides in Hemocytes and Hematopoietic Tissue from Freshwater Crayfish Pacifastacus leniusculus: Characterization and Expression Pattern. Dev. Comp. Immunol. 31(5): 441–455.

Jung, H.J., Y. Park, W.S. Sung, B.K. Suh, J. Lee, K.-S. Hahm and D.G. Lee. 2007. Fungicidal Effect of Pleurocidin by Membrane-active Mechanism and Design of Enantiomeric Analogue for Proteolytic Resistance. Biochim. Biophys. Acta. 1768: 1400–1405.

Kaji, T., N. Sugiyama, S. Ishizaki, Y. Nagashima and K. Shiomi. 2006. Molecular Cloning of Grammistins, Peptide Toxins from the Soapfish Pogonoperca punctata, by Hemolytic Screening of a cDNA Library. Peptides. 27(12): 3069–3076.

Kim, D.-H. and B. Austin. 2008. Characterization of Probiotic Carnobacteria Isolated from Rainbow Trout (Oncorhynchus mykiss) Intestine. Lett. Appl. Microbiol. 47(3): 141–147.

Klaenhammer, T.R. 1993. Genetics of Bacteriocins Produced by Lactic Acid Bacteria. FEMS Microbiol. Rev. 12(1-3): 39–85.

Kobayashi, S., A. Chikushi, S. Tougu, Y. Imura, M. Nishida, Y. Yano and K. Matsuzaki. 2004. Membrane Translocation Mechanism of the Antimicrobial Peptide Buforin 2. Biochemistry. 43(49): 15610–15616.

Kurtz, J. and S.A.O. Armitage. 2006. Alternative Adaptative Immunity in Invertebrates. Trends Immunol. 27: 493–496.

Lauth, X., J.J. Babon, J.A. Stannard, S. Singh, V. Nizet, J.M. Carlberg, V.E. Ostland, M.W. Pennington, R.S. Norton and M.E. Westerman. 2005. Bass Hepcidin Synthesis, Solution Structure, Antimicrobial Activities and Synergism and *in vivo* Hepatic Response to Bacterial Infections. J. Biol. Chem. 280(10): 9272 –9282.

Lauth, X., H. Shike, J.C. Burns, M.E. Westerman, V.E. Ostland, J.M. Carlberg, J.C. Van Olst, V. Nizet, S.W. Taylor, C. Shimizu and P. Bulet. 2002. Discovery and Characterization of Two Isoforms of Moronecidin, a Novel Antimicrobial Peptide from Hybrid Striped Bass. J. Biol. Chem. 277(7): 5030–5039.

Lawniczak, M.K.N., A.I. Barnes, J.R. Linklater, J.M. Boone, S. Wigby and T. Chapman. 2007. Mating and Immunity in Invertebrates. Trends Ecol. Evol. 22: 48–55.

Lee, S.A., Y.K. Kim, S.S. Lim, W.L. Zhu, H. Ko, S.Y. Shin, K.S. Hahm and Y. Kim. 2007. Solution Structure and Cell Selectivity of Piscidin 1 and Its Analogues. Biochemistry. 46(12): 3653–3663.

Lee, S.Y., B.L. Lee and K. Soderhall. 2003. Processing of an Antibacterial Peptide from Hemocyanin of the Freshwater Crayfish Pacifastacus leniusculus. J. Biol. Chem. 278(10): 7927–7933.

Li, C., L. Song, J. Zhao, L. Zhu, H. Zou, H. Zhang, H. Wang and Z. Cai. 2007. Preliminary Study on a Potential Antibacterial Peptide Derived from Histone H2A in Hemocytes of Scallop Chlamys farreri. Fish Shellfish Immunol. 22(6): 663–672.

Lin, H.-J., T.-C. Huang, S. Muthusamy, J.-F. Lee, Y.-F. Duann and C.-H. Lin. 2012. Piscidin-1, an Antimicrobial Peptide from Fish (Hybrid Striped Bass Morone saxatilis x M. chrysops), Induces Apoptotic and Necrotic Activity in HT1080 Cells. Zool. Sci. 29(5): 327–332.

Luders, T., G.A. Birkemo, J. Nissen-Meyer, O. Andersen and I.F. Nes. 2005. Proline Conformation-Dependent Antimicrobial Activity of a Proline-Rich Histone H1 N-Terminal Peptide Fragment Isolated from the Skin Mucus of Atlantic Salmon. Antimicrob. Agents Chemother. 49(6): 2399–2406.

Magnadottir, B. 2006. Innate Immunity of Fish. Fish Shellfish Immunol. 20: 137–151.

Mai, J., W.X.-L. Tian, J.W. Gallant, N. Merkley, Z. Bismas, R.T. Syvitski, S.E. Douglas, J. Ling and Y-H. Li. 2011. A Novel Target-Specific, Salt-Resistant Antimicrobial Peptide Against the Cariogenic Pathogen Streptococcus mutans. Antimicrob. Agents Chemother. 55: 5205–5213.

Marcos, J.F., R.N. Beachy, R.A. Houghten, S.E. Blondelle and E. Pérez-Payá. 1995. Inhibition of a Plant Virus Infection by Analogs of Melittin. Proc. Nat. Acad. Sci. USA. 92(26): 12466–12469.

Mason, A.J., P. Bertani, G. Moulay, A. Marquette, B. Perrone, A.F. Drake, A. Kichler and B. Bechinger. 2007. Membrane Interaction of Chrysophsin-1, a Histidine-Rich Antimicrobial Peptide from Red Sea Bream. Biochemistry. 46(51): 15175–15187.

Mayer, A.M.S. and M.T. Hamann. 2002. Marine Pharmacology in 1999: Compounds with Antibacterial, Anticoagulant, Antifungal, Anthelmintic, Anti-inflammatory, Antiplatelet, Antiprotozoal and Antiviral Activities Affecting the Cardiovascular, Endocrine, Immune and Nervous Systems, and Other Miscellaneous Mechanisms of Action. Comp. Biochem. Physiol., C. 132(3): 315–339.

Mayer, A.M.S. and M.T. Hamann. 2004. Marine Pharmacology in 2000: Marine Compounds with Antibacterial, Anticoagulant, Antifungal, Anti-inflammatory, Antimalarial, Antiplatelet, Antituberculosis, and Antiviral Activities; Affecting the Cardiovascular, Immune, and Nervous Systems and Other Miscellaneous Mechanisms of Action. Mar. Biotechnol. 6(1): 37–52.

Mayer, A.M.S. and M.T. Hamann. 2005. Marine Pharmacology in 2001–2002: Marine Compounds with Anthelmintic, Antibacterial, Anticoagulant, Antidiabetic, Antifungal, Anti-inflammatory, Antimalarial, Antiplatelet, Antiprotozoal, Antituberculosis, and Antiviral Activities; Affecting the Cardiovascular, Immune and Nervous Systems and Other Miscellaneous Mechanisms of Action. Comp. Biochem. Physiol., C. 140(3–4): 265–286.

Mayer, A.M.S., A.D. Rodríguez, R.G.S. Berlinck and M.T. Hamann. 2007. Marine Pharmacology in 2003–4: Marine Compounds with Anthelmintic Antibacterial, Anticoagulant, Antifungal, Anti-inflammatory, Antimalarial, Antiplatelet, Antiprotozoal, Antituberculosis, and Antiviral Activities; Affecting the Cardiovascular, Immune and Nervous Systems, and Other Miscellaneous Mechanisms of Action. Comp. Biochem. Physiol., C. 145(4): 553–581.

Mayer, A.M.S., A.D. Rodríguez, R.G.S. Berlinck and M.T. Hamann. 2009. Marine Pharmacology in 2005–6: Marine Compounds with Anthelmintic, Antibacterial, Anticoagulant, Antifungal, Anti-inflammatory, Antimalarial, Antiprotozoal, Antituberculosis, and Antiviral Activities; Affecting the Cardiovascular, Immune and Nervous Systems, and Other Miscellaneous Mechanisms of Action. Biochim. Biophys. Acta. 1790(5): 283–308.

Mayer, A.M.S., A.D. Rodríguez, R.G.S. Berlinck and N. Fusetani. 2011. Marine Pharmacology in 2007–8: Marine Compounds with Antibacterial, Anticoagulant, Antifungal, Anti-inflammatory, Antimalarial, Antiprotozoal, Antituberculosis, and Antiviral Activities; Affecting the Immune and Nervous System, and Other Miscellaneous Mechanisms of Action. Comp. Biochem. Physiol., C. 153(2): 191–222.

McCall, J.O. and R.K. Sizemore. 1979. Description of a Bacteriocinogenic Plasmid in Beneckea harveyi. Appl. Environ. Microbiol. 38(5): 974–979.

Métivier, A., M.F. Pilet, X. Dousset, O. Sorokine, P. Anglade, M. Zagorec, J.C. Piard, D. Marion, Y. Cenatiempo and C. Fremaux. 1998. Divercin V41, a New Bacteriocin with Two Disulphide Bonds Produced by Carnobacterium divergens V41: Primary Structure and Genomic Organization. Microbiology 144 (Pt 10): 2837–2844.

Metz-Boutigue, M.H., A.E. Kieffer, Y. Goumon and D. Aunis. 2003. Innate Immunity: Involvement of New Neuropeptides. Trends Microbiol. 11(12): 585–592.

Michel-Briand, Y. and C. Baysse. 2002. The Pyocins of Pseudomonas aeruginosa. Biochimie. 84(5-6): 499–510.

Miyasaki, K.T. and R.I. Lehrer. 1998. Beta-sheet Antibiotic Peptides as Potencial Dental Therapeutics. Int. J. Antimicrob. Agents. 9: 269–280.

Montaser, R. and H. Luesch. 2011. Marine Natural Products: a New Wave of Drugs? Future Med. Chem. 3: 1475–1489.

Moon, W.J, D.K. Hwang, E.J. Park, Y.M. Kim and Y.K. Chae. 2007. Recombinant Expression, Isotope Labeling, Refolding, and Purification of an Antimicrobial Peptide, Piscidin. Prot. Express. Purif. 51(2): 141–146.

Nagai, T., T. Osaki and S.-I. Kawabata. 2001. Functional Conversion of Hemocyanin to Phenoloxidase by Horseshoe Crab Antimicrobial Peptides. J. Biol. Chem. 276(29): 27166–27170.

Nayak, S.K. 2010. Role of Gastrointestinal Microbiota in Fish. Aquac. Res. 41(11): 1553–1573.

Nguyen, L.T., E.F. Haney and H.J. Vogel. 2011. The Expanding Scope of Antimicrobial Peptide Structures and Their Modes of Action. Trends Biotechnol. 29(9): 464–472.

Otero-González A.J., B.S. Magalhães, M. Garcia-Villarino, C. López-Abarrategui, D.A. Sousa, S.C. Dias and O.L. Franco. 2010. Antimicrobial Peptides from Marine Invertebrates as a New Frontier for Microbial Infection Control. FASEB J. 24(5): 1320–1334.

Otvos, L. 2000. Antibacterial Peptides Isolated from Insects. J. Pept. Sci. 6: 497–511.

Ovchinnikova, T.V., S.V. Balandin, G.M. Aleshina, A.A. Tagaev, Y.F. Leonova, E.D. Krasnodembsky, A.V. Men'shenin and V.N. Kokryakov. 2006. Aurelin, a Novel Antimicrobial Peptide from Jellyfish Aurelia aurita with Structural Features of Defensins and Channel-blocking Toxins. Biochem. Biophys. Res. Commun. 348(2): 514–523.

Pan, C.-H., J.-Y. Chen, Y.-S.F. Cheng, C.-Y. Chen, I.-H. Ni, J.-F. Sheen, Y.-L. Pan and C. M Kuo. 2007. Gene expression and localization of the Epinecidin-1 antimicrobial peptide in the grouper (Epinephelus coioides), and its role in protecting fish against pathogenic infection. DNA Cell Biol. 26(6): 403–413.

Pan, C.-Y., J.-L. Wu, C.-F. Hui, C.-H. Lin and J.-Y. Chen. 2011. Insights into the Antibacterial and Immunomodulatory Functions of the Antimicrobial Peptide, Epinecidin-1, Against Vibrio vulnificus Infection in Zebrafish. Fish Shellfish Immun. 31(6): 1019–1025.

Park, C.B., M.S. Kim and S.C. Kim. 1996. A Novel Antimicrobial Peptide from Bufo bufo gargarizans. Biochem. Biophys. Res. Commun. 218(1): 408–413.

Park, C.B., J.H. Lee, I.Y. Park, M.S. Kim and S.C. Kim. 1997. A Novel Antimicrobial Peptide from the Loach, Misgurnus anguillicaudatus. FEBS Lett. 411(2-3): 173–178.

Park, C.B., H.S. Kim and S.C. Kim. 1998a. Mechanism of Action of the Antimicrobial Peptide Buforin II: Buforin II Kills Microorganisms by Penetrating the Cell Membrane and Inhibiting Cellular Functions. Biochem. Biophys. Res. Commun. 244: 253–257.

Park, I.Y., C.B. Park, M.S. Kim and S.C. Kim. 1998b. Parasin I, an Antimicrobial Peptide Derived from Histone H2A in the Catfish, Parasilurus Asotus. FEBS Lett. 437(3): 258–262.

Park, K.C, J.A. Osborne, S.C.M. Tsoi, L.L. Brown and S.C. Johnson. 2005. Expressed Sequence Tags Analysis of Atlantic Halibut (Hippoglossus hippoglossus) Liver, Kidney and Spleen Tissues Following Vaccination Against Vibrio anguillarum and Aeromonas salmonicida. Fish Shellfish Immun. 18(5): 393–415.

Park, N.G., U. Silphaduang, H.S. Moon, J.-K. Seo, J. Corrales and E.J. Noga. 2011. Structure—Activity Relationships of Piscidin 4, a Piscine Antimicrobial Peptide. Biochemistry. 50(16): 3288–3299.

Parseghian, M., K.A. Luhrs. 2006. Beyond the Walls of the Nucleus: The Role of Histones in Cellular Signaling and Innate Immunity. Biochem. Cell Biol. 84: 589–604.

Patat, S.A., R.B. Carnegie, C. Kingsbury, P.S. Gross, R. Chapman and K.L. Schey. 2004. Antimicrobial Activity of Histones from Hemocytes of the Pacific White Shrimp. Eur. J. Biochem. 271(23-24): 4825–4833.

Patrzykat, A. and S.E. Douglas. 2005. Antimicrobial Peptides : Cooperative Approach to Protection. Protein Pept. Lett. 12: 19–25.

Patrzykat, A. and S.E. Douglas. 2003. Gone Gene Fishing: How to Catch Novel Marine Antimicrobials. Trends Biotechnol. 21(8): 362–369.

Patrzykat, A., J.W. Gallant, J.-K. Seo, J. Pytyck and S.E. Douglas. 2003. Novel Antimicrobial Peptides from Flatfish Genes. Antimicrob. Agents Chemother. 47: 2464–2470.

Pazgier, M., D. Hoover, D. Yang, W. Lu and J. Lubkowski. 2006. Human B-defensins. Cell. Mol. Life Sci. 63(11): 1294–1313.

Pelegrini, P.B. and O.L. Franco. 2005. Plant γ-thionins: Novel Insights on the Mechanism of Action of a Multi-functional Class of Defense Proteins. Int. J. Biochem. Cell Biol. 37(11): 2239–2253.

Pérez-Sánchez, T., J.L. Balcázar, Y. García, N. Halaihel, D. Vendrell, I. de Blas, D.L. Merrifield and I. Ruiz-Zarzuela. 2011. Identification and Characterization of Lactic Acid Bacteria Isolated from Rainbow Trout, Oncorhynchus mykiss (Walbaum), with Inhibitory Activity Against Lactococcus garvieae. J. Fish Dis. 34(7): 499–507.

Pilet, M.-F., X. Dousset, R. Barré, G. Novel, M. Desmazeaud and J-C. Piard. 1995. Evidence for Two Bacteriocins Produced by Carnobacterium piscicola and Carnobacterium divergens Isolated from Fish and Active Against Listeria monocytogenes. J. Food Prot. 58(3): 256–262.

Porcelli, F., B. Buck, D.-K. Lee, K.J. Hallock, A. Ramamoorthy and G. Veglia. 2004. Structure and Orientation of Pardaxin Determined by NMR Experiments in Model Membranes. J. Biol. Chem. 279(44): 45815–45823.

Prasad, S., P.C. Morris, R. Hansen, P.G. Meaden and B. Austin. 2005. A Novel Bacteriocin-like Substance (BLIS) from a Pathogenic Strain of Vibrio harveyi. Microbiology. 151(9): 3051–3058.

Rajanbabu, V. and J.-Y. Chen. 2011. Applications of Antimicrobial Peptides from Fish and Perspectives for the Future. Peptides. 32(2): 415–420.

Reddy, K.V.R, R.D. Yedery and C. Aranha. 2004. Antimicrobial Peptides : Premises and Promises. Int. J. Antimicrob. Agents. 6: 536–547.

Ren, H.-L., K.-J. Wang, H.-L. Zhou and M. Yang. 2006. Cloning and Organisation Analysis of a Hepcidin-like Gene and cDNA from Japan Sea Bass, Lateolabrax japonicus. Fish Shellfish Immun. 21(3): 221–227.

Richards, R.C., D.B. O'Neil, P. Thibault and K.V. Ewart. 2001. Histone H1: An Antimicrobial Protein of Atlantic Salmon (Salmo salar). Biochem. Biophys. Res. Commun. 284(3): 549–555.

Rihakova, J., Y. Belguesmia, V.W. Petit, M.-F. Pilet, H. Prévost, X. Dousset and D. Drider. 2009. Divercin V41 from Gene Characterization to Food Applications: 1998-2008, a Decade of Solved and Unsolved Questions. Lett. Appl. Microbiol. 48(1): 1–7.

Riley, M.A. and J.E. Wertz. 2002. Bacteriocins: Evolution, Ecology, and Application. Ann. Rev. Microbiol. 56(1): 117–137.

Rinaldi, A.C. 2002. Antimicrobial Peptides from Amphibian Skin: An Expanding Scenario: Commentary. Curr. Op. Chem. Biol. 6(6): 799–804.

Ringø, E. and F.-J. Gatesoupe. 1998. Lactic Acid Bacteria in Fish: a Review. Aquaculture. 160(3-4): 177–203.

Rodrigues, P.N.S., S. Vázquez-Dorado, J.V. Neves and J.M. Wilson. 2006. Dual Function of Fish Hepcidin: Response to Experimental Iron Overload and Bacterial Infection in Sea Bass (Dicentrarchus labrax). Dev. Comp. Immunol. 30(12): 1156–1167.

Ruangsri, J., J.M.O. Fernandes, J.H.W.M. Rombout, M. Brinchmann and V. Kiron. 2012. Ubiquitous Presence of Piscidin-1 in Atlantic Cod as Evidenced by Immunolocalisation. BMC Vet. Res. 8(1): 46.

Rydlo, T., J. Miltz and A. Mor. 2006. Eukaryotic Antimicrobial Peptides: Promises and Premises in Food Safety. J. Food Sci. 71: 125–135.

Saint, N., H. Cadiou, Y. Bessin and G. Molle. 2002. Antibacterial Peptide Pleurocidin Forms Ion Channels in Planar Lipid Bilayers. Biochim. Biophys. Acta. 1564(2): 359.

Salerno, G., N. Parrinello, P. Roch and M. Cammarata. 2007. cDNA Sequence and Tissue Expression of an Antimicrobial Peptide, Dicentracin; a New Component of the Moronecidin Family Isolated from Head Kidney Leukocytes of Sea Bass, Dicentrarchus labrax. Comp. Biochem. Physiol. 146(4): 521–529.

Salzet, M. 2005. Neuropeptide-Derived Antimicrobial Peptides from Invertebrates for Biomedical Applications. Curr. Med. Chem. 12(26): 3055–3061.

Satish Kumar, R., P. Kanmani, N. Yuvaraj, K.A. Paari, V. Pattukumar and V. Arul. 2011. Purification and characterization of enterocin MC13 produced by a potential aquaculture probiont Enterococcus faecium MC13 isolated from the gut of Mugil cephalus. Can. J. Microbiol. 57(12): 993–1001.

Sequeiros, C., M. Vallejo, E. Marguet and N. Olivera. 2010. Inhibitory Activity Against the Fish Pathogen Lactococcus garvieae Produced by Lactococcus lactis TW34, a Lactic Acid Bacterium Isolated from the Intestinal Tract of a Patagonian Fish. Arch. Microbiol. 192(4): 237–245.

Shai, Y. 2002. Mode of Action of Membrane Active Antimicrobial Peptides. Biopolymers. 66: 236–248.

Shehane, S.D. and R.K. Sizemore. 2002. Isolation and Preliminary Characterization of Bacteriocins Produced by Vibrio vulnificus. J. Appl. Microbiol. 92(2): 322–328.

Shi, J. and A.C. Camus. 2006. Hepcidins in Amphibians and Fishes: Antimicrobial Peptides or Iron-regulatory Hormones? Dev. Comp. Immunol. 30(9): 746–755.

Shike, H., X. Lauth, M.E. Westerman, V.E. Ostland, J.M. Carlberg, J.C. Van Olst, C. Shimizu, P. Bulet and J.C. Burns. 2002. Bass Hepcidin Is a Novel Antimicrobial Peptide Induced by Bacterial Challenge. Eur. J. Biochem. 269(8): 2232–2237.

Shike, H., C. Shimizu, X. Lauth and J.C. Burns. 2004. Organization and Expression Analysis of the Zebrafish Hepcidin Gene, an Antimicrobial Peptide Gene Conserved Among Vertebrates. Dev. Comp. Immunol. 28(7–8): 747–754.

Shinnar, A.E., K.L. Butler and H.J. Park. 2003. Cathelicidin Family of Antimicrobial Peptides: Proteolytic Processing and Protease Resistance. Bioorg. Chem. 31(6): 425–436.

Silphaduang, U. and E.J. Noga. 2001. Antimicrobials: Peptide Antibiotics in Mast Cells of Fish. Nature. 414(6861): 268–269.

Silva, F.C.P., M.F.G. Brito, L.M. Farias and J.R. Nicoli. 2005. Composition and Antagonistic Activity of the Indigenous Intestinal Microbiota of Prochilodus argenteus Agassiz. J. Fish Biol. 67(6): 1686–1698.

Sitaram, A. 2006. Antimicrobial Peptides with Unusual Amino Acid Composition and Unusual Structures. Curr. Med. Chem. 13: 763–771.

Smith, V.J., A.P. Desbois and E.A. Dyrynda. 2010. Conventional and Unconventional Antimicrobials from Fish, Marine Invertebrates and Micro-algae. Mar. Drugs. 8(4): 1213–1262.

Spanggaard, B., I. Huber, J. Nielsen, E.B. Sick, C.B. Pipper, T. Martinussen, W.J. Slierendrecht and L. Gram. 2001. The Probiotic Potential Against Vibriosis of the Indigenous Microflora of Rainbow Trout. Environ. Microbiol. 3(12): 755–765.

Sperstad, S.V., T. Haug, H.-M. Blencke, O.B. Styrvold, C. Li and K. Stensvåg. 2011. Antimicrobial Peptides from Marine Invertebrates: Challenges and Perspectives in Marine Antimicrobial Peptide Discovery. Biotechnol. Adv. 29(5): 519–530.

Stec, B. 2006. Plant Thionins—the Structural Perspective. Cell. Mol. Life Sci. 63: 1370–1385.

Steiner, H., D. Hultmark, A. Engstrom, H. Bennich and H.G. Boman. 1981. Sequence and Specificity of Two Antibacterial Proteins Involved in Insect Immunity. Nature. 16: 246–248.

Stoffels, G., A. Gudmundsdottir and T. Abee. 1994. Membrane-associated proteins encoded by the nisin gene cluster may function as a receptor for the lantibiotic carnocin UI49. Microbiology. 140(6): 1443–1450.

Stoffels, G., J. Nissen-Meyer, A. Gudmundsdottir, K. Sletten, H. Holo and I.F. Nes. 1992. Purification and Characterization of a New Bacteriocin Isolated from a Carnobacterium sp. Appl.Environ. Microbiol. 58(5): 1417–1422.

Sugiarto, H. and P.-L. Yu. 2004. Avian Antimicrobial Peptides: The Defense Role of β-defensins. Biochem. Biophys. Res. Commun. 323(3): 721–727.

Sugita, H., N. Matsuo, Y. Hirose, M. Iwato and Y. Deguchi. 1997. *Vibrio* sp. Strain NM 10, Isolated from the Intestine of a Japanese Coastal Fish, has an Inhibitory Effect Against Pasteurella piscicida. Appl. Environ. Microbiol. 63(12): 4986–4989.

Sugiyama, N., M. Araki, M. Ishida, Y. Nagashima and K. Shiomi. 2005. Further Isolation and Characterization of Grammistins from the Skin Secretion of the Soapfish Grammistes sexlineatus. Toxicon. 45(5): 595–601.

Suttle, C.A. 2005. Viruses in the Sea. Nature. 437(7057): 356–361.

Syvitski, R.T., I. Burton, N.R. Mattatall, S.E. Douglas and D.L. Jakeman. 2005. Structural Characterization of the Antimicrobial Peptide Pleurocidin from Winter Flounder. Biochemistry. 44(19): 7282–7293.

Tennessen, J.A. 2005. Molecular Evolution of Animal Antimicrobial Peptides: Wisespread Moderate Positive Selection. J. Evol. Biol. 18: 1387–1394.

Tincu, J.A., L.P. Menzel, R. Azimov, J. Sands, T. Hong, A.J. Waring, S.W. Taylor and R.I. Lehrer. 2003. Plicatamide, an Antimicrobial Octapeptide from Styela plicata Hemocytes. J. Biol. Chem. 278(15): 13546–13553.

Tincu, J.A. and S.W. Taylor. 2004. Antimicrobial Peptides from Marine Invertebrates. Antimicrob. Agents Chemother. 48(10): 3645–3654.

Tota, B., A.M. Quintieri, V. Di Felice and M.C. Cerra. 2007. New Biological Aspects of Chromogranin A-derived Peptides: Focus on Vasostatins. Comp. Biochem. Physiol. A Mol. Integr. Physiol. 147: 11–18.

Ulvatne, H. 2003. Antimicrobial Peptides: Potencial Use in Skin Infections. Am. J. Clin. Dermatol. 4: 591–595.

van Holde, K.E., K.I. Miller and H. Decker. 2001. Hemocyanins and Invertebrate Evolution. J. Biol. Chem. 276: 15563–15566.

Verschuere, L., G. Rombaut, P. Sorgeloos and W. Verstraete. 2000. Probiotic Bacteria as Biological Control Agents in Aquaculture. Microbiol. Mol. Biol. Rev. 64(4): 655–671.

Wachinger, M., A. Kleinschmidt, D. Winder, N. von Pechmann, A. Ludvigsen, M. Neumann, R. Holle, B. Salmons, V. Erfle and R. Brack-Werner. 1998. Antimicrobial Peptides Melittin and Cecropin Inhibit Replication of Human Immunodeficiency Virus 1 by Suppressing Viral Gene Expression. J. Gen. Virol. 79(4): 731–740.

Wang, K.F., R. Nagarajan, C.M. Mello and T.A. Camesano. 2011. Characterization of Supported Lipid Bilayer Disruption By Chrysophsin-3 Using QCM-D. J. Phys. Chem. B. 115(51): 15228–15235.

Wang, K.-J., W.-S. Huang, M. Yang, H.-Y. Chen, J. Bo, S.-J. Li and G.-Z. Wang. 2007. A Male-specific Expression Gene, Encodes a Novel Anionic Antimicrobial Peptide, Scygonadin, in Scylla serrata. Mol. Immunol. 44(8): 1961–1968.

Wang, W., A.M. Cole, T. Hong, A.J. Waring and R.I. Lehrer. 2003. Retrocyclin, an Antiretroviral θ-Defensin, Is a Lectin. J. Immunol. 170(9): 4708–4716.

Wiesner, J., A. Vilcinskas. 2010. Antimicrobial Peptides: The Ancient Arm of the Human Immune System. Virulence. 1(2150-5594): 440–464.

Yasin, B., M. Pang, J.S. Turner, Y. Cho, N.N. Dinh, A.J. Waring, R.I. Lehrer and E.A. Wagar. 2000. Evaluation of the Inactivation of Infectious Herpes Simplex Virus by Host-Defense Peptides. Eur. J. Clin. Microbiol. Infect. Dis. 19(3): 187.

Yeaman, M.R. and N.Y. Yount. 2003. Mechanisms of Antimicrobial Peptide Action and Resistance. Pharmacol Rev. 55(1): 27–55.

Zai, A.S., S. Ahmad and S.A.O. Armitage. 2009. Bacteriocin Production by Indigenous Marine Catfish Associated *Vibrio* spp. Pak. J. Pharma. Sci. 22(2): 162–167.

Zaiou, M. 2007. Multifunctional Antimicrobial Peptides: Therapeutic Targets in Several Human Diseases. J. Mol. Med. 85: 317–29.

Zasloff, M. 1987. Magainins, a Class of Antimicrobial Peptides from Xenopus Skin: Isolation, Characterization of Two Active Forms, and Partial cDNA Sequence of a Precursor. Proc. Natl. Acad. Sci. USA. 84: 5449–5453.

Zasloff, M. 2002. Antimicrobial Peptides of Multicellular Organisms. Nature. 415: 389–395.

Zhang, X., C. Huang and Q. Qin. 2004a. Antiviral Properties of Hemocyanin Isolated from Shrimp Penaeus Monodon. Antiviral Res. 61(2): 93–99.

Zhang, Y-A., J. Zou, C.-I. Chang and C.J. Secombes. 2004b. Discovery and Characterization of Two Types of Liver-expressed Antimicrobial Peptide 2 (LEAP-2) Genes in Rainbow Trout. Vet. Immunol. Immunopathol. 101(3–4): 259–269.
Zhou, J.-G., J.-G. Wei, D. Xu, H.-C. Cui, Y. Yan, Z.-L. Ou-Yang, X.-H. Huang, Y.-H. Huang and Q.-W. Qin. 2011. Molecular Cloning and Characterization of Two Novel Hepcidins from Orange-spotted Grouper, Epinephelus coioides. Fish Shellfish Immun. 30(2): 559–568.
Zweifel, U.L. and A. Hagstrom. 1995. Total Counts of Marine Bacteria Include a Large Fraction of Non-Nucleoid-Containing Bacteria (Ghosts). Appl. Environ. Microbiol. 61(6): 2180–2185.

6

Secretagogues Effects of Fish and Shellfish By-Products Upgrading

Rozenn Ravallec, Benoit Cudennec* and *Pascal Dhulster*

ABSTRACT

For many years, numerous kinds of processes based on enzymatic hydrolysis at various pH levels, involving added plant or bacterial enzymes and after inactivation by heating of endogenous enzymes present in the raw material, or alternatively based on the action of endogenous enzymes, have contributed to the degradation of marine by-products proteins in order to produce fractions exerting secretagogue activities in various domains. The action of secretion covers a wide range of biological activities ranging from the regulation of the digestive enzymes and of satiating hormones to the stimulation of growth, and present an important aspect in the development of aquaculture.

INTRODUCTION

The search for new molecules in fish protein hydrolysates is of great interest in animal feeding as it falls in the aquaculture, fertilizer, cosmetic, and pharmacologic sector. Numerous studies of alternative sources of

Laboratoire de PROcédés BIOlogiques, Génie Enzymatique et Microbien—ProBioGEM, Boulevard Paul Langevin, 59650 Villeneuve d'Ascq, France.

Emails: rozenn.ravallec-ple@univ-lille1.fr; benoit.cudennec@univ-lille1.fr; pascal.dhulster@univ-lille1.fr

* Corresponding author

fish and shellfish meal have been conducted because of the worldwide expansion of aquaculture and demands on feed (Kihara et al. 2011, Tidwell and Allan 2001). High added value fish protein hydrolysate (FPH) can be commercially produced from fish by-products through simple engineering (Chakraborty and Madravan 1977, Kim et al. 2000) and can subsequently be used as food supplements (Mackie 1982, Durand and Lagoin 1983, Merritt 1982), animal feed or fertilizer (Venugopal 1994). The different processes of hydrolysis or autolysis of animal proteins generate molecules ranging from individual aminoacids to peptides of various sizes, some of which are likely to harbour biological properties of economic interest in the development of food supplements (Hsueh and Moskowitz 1973) or in aquaculture such as growth factors or secretagogues. The process generating these peptides from the original raw material, requires control and optimization to obtain the most reproducible and biologically active hydrolysates. Changes in the different parameters such as time or temperature conditions of hydrolysis, as well as the enzyme/substrate ratio will generate hydrolysates composed of differently sized molecules presenting different biochemical properties (Quaglia and Orban 1987, Cudennec et al. 2012).

Growth factors are defined as a group of polypeptide regulatory agents which act to control a host of cellular responses (cell multiplication or DNA synthesis) by a series of mechanisms analogous to that of classical endocrine hormones. Secretagogue molecules such as gastrin and cholecystokinin exhibit a wide spectrum of activities ranging from the stimulation of protein synthesis to the control of intestinal mobility and the secretion of digestive enzymes. Molecules harbouring structural and/or functional characteristics of gastrin or growth factors have already been detected in fish hydrolysates (Cancre et al. 1999). Igarashi et al. 2005 reported that amino acids or peptides stimulated pepsin secretion *in vivo*.

Immunomodulators are a group of biomolecules able to enhance the phagocytosis and increase the production of reactive oxygen metabolites in macrophages. The ability of macrophages to kill pathogens is of primary importance in the prevention of diseases, by the activity of immunomodulators, and could be one of the ways to enhance the nonspecific immune defense system (Gildberg et al. 1996).

The regulation of maintenance of body weight is integrated by several endocrinal signals such as CCK (cholecystokinins) and GLP-1 (glucagon-like peptide-1), particularly involved in the short-term response related to meal size. CCK was the first intestinal hormone evidenced in the regulation of food intake and is one of the most studied satiety peptide hormone (Gibbs et al. 1973). CCK is mainly synthetized in two forms (CCK-8 and CCK-33) in response to a meal stimulation from I-cells localized in the duodenum. It is also synthesized in the central nervous system (Beinfeld 1995, Liddle 1997). Intestinal CCK release leads to several gastrointestinal actions such as intestinal motility, gall bladder contraction, inhibition of gastric emptying and stimulation of pancreatic secretion (Strader and Woods 2005).

This chapter will cover a wide group of secretagogues activities from the regulation of the enzyme secretion to the study of feed regulating hormones secretion. Using specific bioassays, the authors detected a series of biological activities related to the above-mentioned factors in fish and shellfish hydrolysates: gastrin/CCK, growth factors, immunomodulating and food intake regulation peptides. The characterization of these peptides in marine hydrolysates could be of importance because of the development of aquaculture, which requires strict control of feed quality and ingredients sources.

GROWTH STIMULATION

Fish protein hydrolysates have been used as substitutes for fish meal in aquaculture feed, in order to enhance the growth and survival of marine fish (Kotzamanis et al. 2007). Growth factor are defined as a diverse group of polypeptide regulatory agents acting to control a host of cellular response, such as cell multiplication, by a series of mechanisms analogous to those of classic endocrine hormones. Cancre et al. demonstrated that different sources of hydrolysates such as shrimp waste (*Pandalus borealis*), cod (*Gadus morhua*) head, and head and viscera of sardine (*Sardina pilchardus*), obtained after hydrolysis or autolysis, were tested on fibroblast cell cultures 3T3. The level of hydrolysis seems to play an important role in the presence of biological peptides. Elution profile on a gel filtration Sephadex G-50 column was used to estimate the degree of hydrolysis of the fractions studied. Growth factor-like activities were found in less-hydrolyzed fractions. An interesting growth factor-like activity (150% of growth/against the control) in sardine by-products was also observed after only two hours of the hydrolysis by the Alacalase® (Ravallec-Plé et al. 2001). The authors explained that growth factor activities were detected in the hydrolysates with lower degree of hydrolysis. Thus, to obtain biological properties, the most important factor is not an extended process, but protein's degradation sufficient for the production of bioactive peptides. Ravallec-Plé et al. in 2000 also demonstrated the presence of growth factor-like activity in cod (*Gadus morhua*) muscle hydrolysates. Under particular conditions, the authors obtained 160% stimulation of cell growth (3T3) with only 25 $\mu g\ ml^{-1}$ hydrolysates and concluded that the process of hydrolysis plays a key factor by extending the time of protein degradation in generating these molecules.

The incorporation of FPH in artificial larval diets has been suggested as an alternative approach for overcoming the limited digestive capacity of sea bass (*Dicentrarchus labrax*) larvae (Kotzamanis et al. 2007). In this study, the results indicate that different molecular weight fractions and concentrations of feed-soluble peptides (from whole sardine hydrolysate, *Sardina pilchardus*) may affect the growth performance and immunological status of fish larvae.

Nguyen et al. 2012, conducted a feeding trial to evaluate the effect of the supplementation of hydrolysates from tuna head on the survival and growth of *Penaeus vannamei*. After six weeks of experiment, the survival rate (the ratio between the number of living shrimps at the end and the beginning of the experiment), weight gain, feed conversion ratio and protein efficiency ratio were compared between the different groups. The results showed that the diets supplemented with the hydrolysates of tuna head improved significantly the four zootechnical parameters studied. On the contrary, authors concluded that the survival rates and growth performances for the groups fed with the diets containing a mixture of protein powders were not significantly different or even lower compared to the control. All these results confirm the possibility to promote a better utilization of fishing sources, by employing fish by-products as protein sources, in fish or shellfish meal in aquaculture.

ENZYME SECRETION

In addition to the benefits in growth performance in *in vivo* experiments on fish larvae, it has been shown that fractions of peptides derived from hydrolysed muscle of cod (*Gadus morhua*) stimulate the production of digestive enzymes expression and cholecystokinin contents (Cahu et al. 2004). Secretagogues molecules like gastrin and cholecystokinines (CCK) exhibit a large spectrum of activities from the stimulation of protein synthesis to the aim of this section, the secretion of digestive enzymes (Beinfeld 1995). Cancre et al. used different sources of hydrolysates such as shrimp waste (*Pandalus borealis*), cod (*Gadus morhua*) head, and head and viscera of sardine (*Sardina pilchardus*), obtained after hydrolysis or autolysis and tested by gastrin radioimmunoassay. They show that the most-hydrolyzed fraction contains small peptides immunogically related to gastrin and cholecystokinin. The same kind of results was also presented by Ravallec-Plé et al. 2000. In this publication, several fractions were shown to contain biologically active peptides such as secretagogues (gastrin and cholecystokinin). Under optimum conditions they obtained about 0.25 pg gastrin-like peptides per mg of hydrolysate. All these results show how the process of hydrolysis plays a key factor by extending the time of protein degradation in generating these active molecules.

Extended researches were performed by Ravallec-Plé and van Wormhoudt, in 2003. In this work, peptides with gastrin immunoreactivity were measured in cod muscle (*Gadus morhua*) and shrimp heads

(*Penaeus aztecus*) extracts and alcalase hydrolysates and separated by two chromatographic steps. The secretagogue activities present in crude extracts fractions were examined with or without specific antagonists of CCK receptors in AR4-2J cells, a rat pancreatic acinar cell line able to secrete amylase after the binding of gastrin/CCK related molecules on CCK receptors (CCK1 and CCK2). Several sub-fractions significantly stimulate amylase release, up to 110%. These stimulatory effects could be completely inhibited by the presence of L 365, 260 specific synthetic antagonist of the CCK2 receptors (also present in the gastrointestinal tract). After hydrolysis of the raw material, the samples were partially fractionated by two chromatographic steps and potential active fractions detected by a gastrin-CCK radioimmunoassay. The molecular weight of the active fractions were lower than for the extracts. Stimulation of amylase release was higher than with extracts, and the inhibition by L 365, 260, less pronounced. These results show that some peptides remaining after hydrolysis or extraction still exert biological activities and have to be tested in nutritional studies (Table 1).

In vivo, Cahu et al. suggested that dietary protein level and chain length combined with protein intraluminal proteolytic activity regulate the CCK level in sea bass (*Dicentrarchus labrax*) larvae as in other vertebrates since the lowest CCK level was found in the group fed with the highest protein hydrolysate level. The CCK concentration assayed in larvae fed diets with a low protein content or different hydrolysates levels is compatible with the existence of an indirect mechanism controlling CCK release and mediating pancreatic enzyme secretion (Cahu et al. 2004).

Table 1 Amount of gastrin equivalent for each fraction after RadioImmunoAssay RIA (from: Ravallec-Plé and Van Wormhoudt 2003).

Source	*Fractions (Da)*	*Pg of gastrin equivalent/mg of dry mass*
Shrimp extraction	A: >20,000	1.00
	B: 2000–500	32.37
	D: <500	1.58
Shrimp hydrolysate	A: >20,000	3.80
	B: 4000–1000	4.00
	B: 1000–500	2.50
	D: <500	1.12
Cod extraction	A: >20,000	-
	B: 10 000–3500	-
	B: 1500–500	9.3
	D: <500	0.488
Cod hydrolysate	A: >20,000	0.21
	B: 4000–1000	1.04
	B: 1000–500	1.32
	D: <500	0.08

IMMUNOMODULATING ACTIVITY

In the last years, an increasing interest in the development of immunomodulators from natural sources was observed with the application in several domains such as diet supplementation for aquaculture. This research area represents a modern and promising tool in aquaculture, enhancing the resistance of cultured fish to disease and stress (Bagni et al. 2000, Duarte et al. 2006) by improving the systemic immune response or by modulating the functions of immunocompetent cells (Perdigon et al. 1986). Duarte et al. studied the effect of a commercial fish protein concentrate (FPC), obtained by the fermentation of fish protein by yeast and show that different accumulative doses of FPC are able to induce the release of cytokines (activating the immune cells distant from the site of induction of the immune response) without inducing any inflammatory immune response in BALB/c female mice (Duarte et al. 2006).

Hou et al. 2012, identified immunomodulating peptides from enzymatic hydrolysates by trypsin of Alaska Pollock frame following the lymphocyte proliferation rate (Table 2). Although the immunomodulating mechanism of these bioactives peptides remains uncertain, studies confirm that immunomodulatory activity was related to lymphocyte proliferation and antibody synthesis (Yang et al. 2009, Gauthier at al. 2006). Literature show that peptides, generated in a fish muscle hydrolysate, and involved in the stimulation of oxidative burst reactions in Atlantic salmon (*Salmo salar*) head kidney leucocytes, are medium size peptides (3000 Da > Mw > 500 Da) (Gildberg et al. 1996, Bogwald et al. 1996).

All these researches provided a scientific basis for the preparation of immunomodulating food from fish by-products with a demonstrated capacity to enhance non-specific host defense mechanisms.

Table 2 Immunomodulating peptides from enzymatic hydrolysates of Alaska Pollock frame (adapted from: Hou et al. 2012).

Sequence	*Molecular weights (Da)*	*Activity (Lymphocyte proliferation rate)*
Asn-Gly-Met-Thr-Tyr	584	35.92%
Asn-Gly-Leu-Ala-Pro	470	32.96%
Trp-Thr	305	31.55%

FEED REGULATING HORMONES SECRETION

In the literature, authors previously described that FPH could harbour secretagogues activities (Ravallec-Plé et al. 2000, Guerard and Ravallec-Plé 2001, Ravallec-Plé et al. 2001, Ravallec-Plé and Van Wormhoudt 2003, Picot et al. 2010) particularly involved in the stimulation of feed of food regulating hormones secretion (Cudennec et al. 2008, Bougatef et al. 2010, Cudennec et al[b]. 2012).

Excess of food intake largely contributes to body weight increase and obesity that predisposes to numerous pathologies such as diabetes and hypertension (Arora and Anubhuti 2006). The long-term regulation of maintenance of body weight is integrated by endocrine signals and coupled with short-term signals related to meal size, like cholecystokinin (CCK) and GLP-1. This contributes to regulate the daily energy intake in synchronization with long-term body weight regulation (Arora and Anubhuti 2006, Schwartz et al. 2000). Among the individual alimentary compounds, protein and their hydrolysis products (peptides of various sizes) appear to be the main stimulant of the intestinal CCK secretion (Liddle 1997).

Indeed, it has been demonstrated that protein hydrolysates are able to stimulate CCK and GLP-1 secretions and their respective gene transcription in the STC-1 endocrine cell line (Cordier-Bussat et al. 1997). Because of the strong potential of protein hydrolysates to stimulate satietogenic peptides secretion, their utilization as appetite suppressive products in the prevention and treatment of obesity has evinced a growing interest last few years.

With an aim to upgrade fish by-products and to find new appetite suppressing peptides from fish sources, authors demonstrated using *in vitro* models that industrially produced hydrolysates from blue whiting muscle were able to stimulate the secretion of CCK and GLP-1 in the enteroendocrine STC-1 cell line (Cudennec et al. 2008, Cudennec et al.[a] 2012), as shown in Fig. 1. These stimulating effects were dose dependant for both CCK and GLP-1 secretions and control experiments performed using undigested protein (BSA) and amino-acid mixture demonstrated that they were mainly due to peptide molecules.

In a second approach, in order to demonstrate that the *in vitro* effect observed on the secretion of CCK and GLP-1 could be related to satietogenic properties of this fish hydrolysate, Cudennec at al. investigated the effect of oral administration of blue whiting muscle hydrolysate (BWMH) on food intake and associated physiological processes in rats. Thus, the effect of BWMH preloads administration at different concentrations on long- and short-term food intake were studied and demonstrated a dose-dependent decrease of short-term food intake. Indeed, at the end of the two weeks of experimentation, FPH generated a reduction in the increase of body weight (Fig. 2).

In addition, after a 24 hour fasting period and re-administation of preloads, blood sample was taken in order to determine plasmatic markers of energy homeostasis (Fig. 3). The results obtained demonstrated a relationship between experiments performed *in vitro* and *in vivo*. The plasmatic levels of CCK and active GLP-1 dose-dependent increase observed

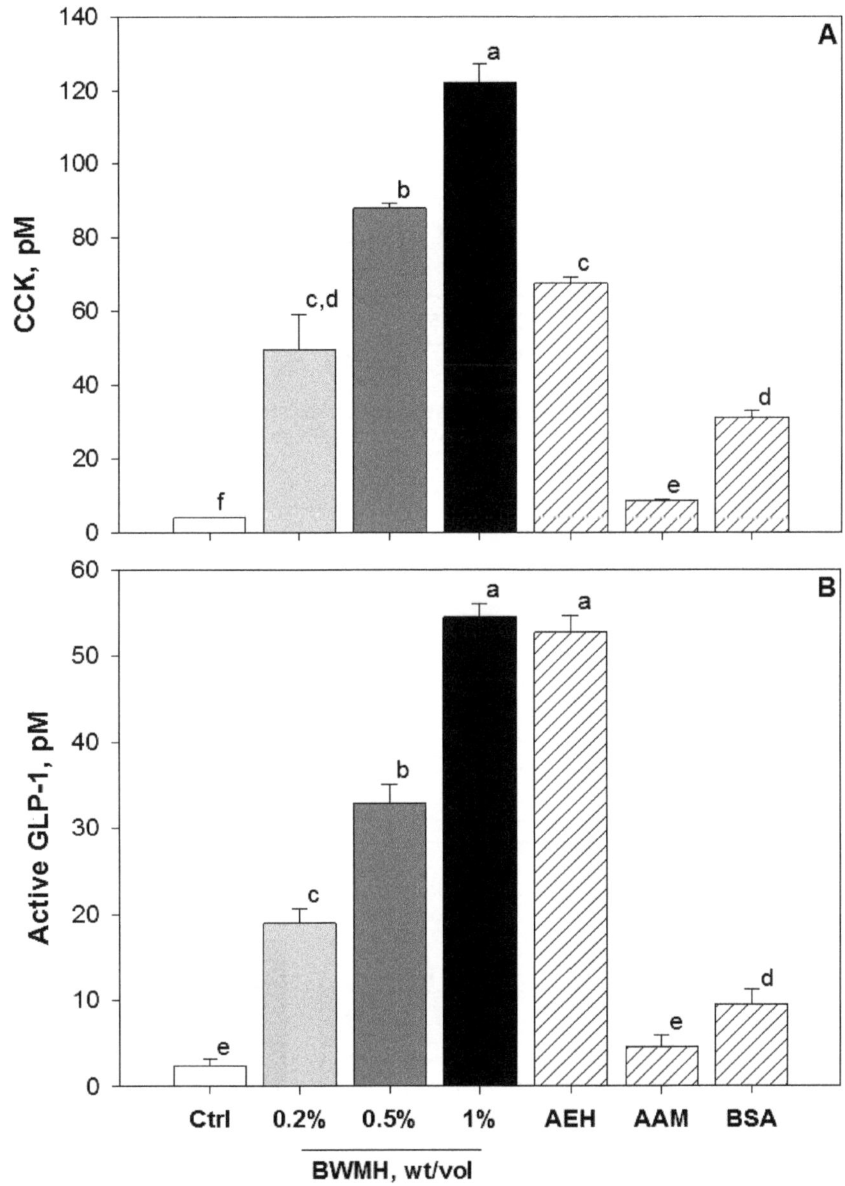

Fig. 1 Effect of BWMH on CCK and active GLP-1 releasing activity of STC-1 cells. BWMH was tested at different concentrations (0.2 to 1.0%, wt/vol) on CCK (panel A) and active GLP-1 (panel B) release and compared to those obtained with albumin egg hydrolysate (AEH, a positive control), bovine serum albumin (BSA) and free amino acid mixture (AAM), each at a 1% wt/vol concentration. Values are means ± SEM of four repeated measurements. Means without a common letter are different ($p<0.05$) (Cudennec et al. 2012a).

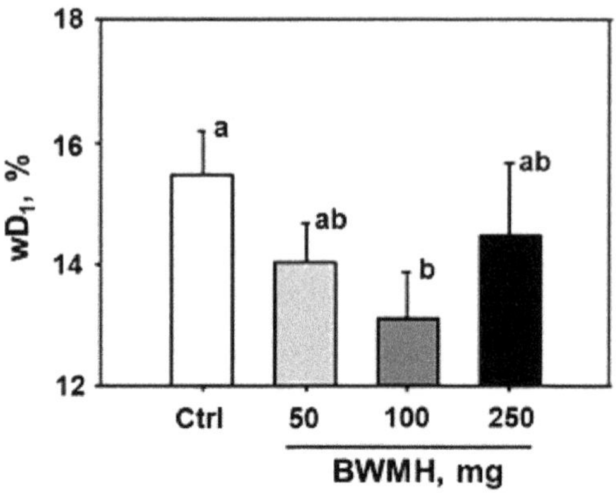

Fig. 2 Body weight increase was expressed as the percentage of rat weight at D_1 (wD_1), values presented are intra-group means (n=8, ± SEM). Means without a common letter are different ($p<0.05$) (Cudennec et al. 2012a).

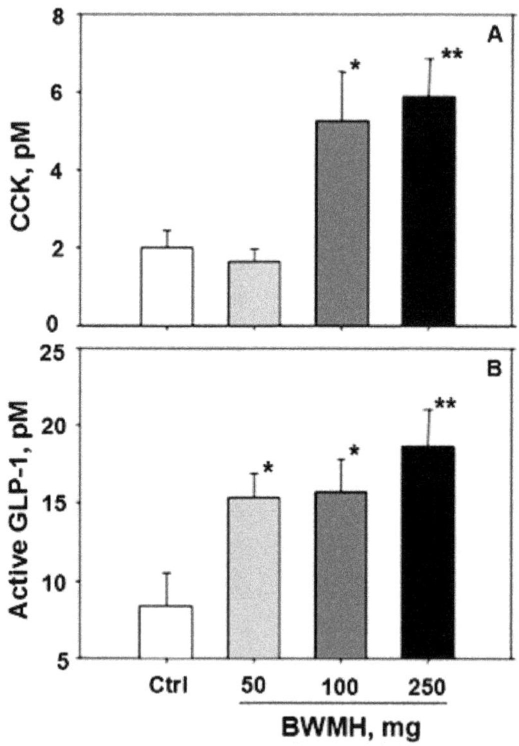

Fig. 3 Effect of BWMH on plasmatic GLP-1 and CCK levels of energetic metabolism, after a 24h fasting period and assayed 30 minutes after preload oral administration. Plasmatic markers: CCK (A), active GLP-1 (B). Values are intra-group means (n=8 ± SEM) of three measurements for each rat. Differences between the control group and 50, 100 and 250 mg groups are indicated by *: $p<0.05$ and **: $p<0.01$ (Cudennec et al. 2012a).

after administration of BWMH is related to the reduction of the short-term food intake observed. The same kind of results were observed by Bougatef et al. 2010, and revealed a possible mechanism of the beneficial effects of FPH in appetite regulation, and this might be important to prevent the risk of a number of medical conditions including obesity or type II diabetes. *In vitro* and *in vivo* experiments show the possibility of using FPH in nutraceutical ingredients to regulate the food intake, and in consequence the body weight loss or gain.

CONCLUSION

This chapter shows biological activities from hydrolysates from different marine species. There is no general rule by which to determine the best hydrolysate with the most important activity. Each raw material, each enzyme, and each conditions of hydrolysis will yield different final peptides. However, this summary mainly provides evidence of the peptides distribution and activities in compound diets.

The relationship between the sequence and the activity is always studied and represents a fundamental research for the comprehension of the complex mechanism of biological activities. In this context, another point to highlight is the necessity of a better understanding of the *in vivo* experiments to confirm and understand the mechanisms involved in the biological activity. These hydrolysates represent an important source of sequences, able to be generated after the hydrolysis of marine proteins by gastrointestinal enzymes (Bourseau et al. 2009, Cudennec et al. 2012a). From an upgrading point of view, an industrial scale fraction exhibiting an activity able to enhance growth, digestion, survival and feed regulation of larvae or fish species would be of great importance in several domains such as feed or aquaculture. Further research is required to characterize, purify and identify the peptides involved in the biological activities and to determine their optimal inclusion levels in diets for fish and shellfish species.

REFERENCES

Arora, S., Anubhuti 2006. Role of neuropeptides in appetite regulation and obesity—a review. Neuropeptides 40: 375–401.

Bagni, M., L. Archetti, M. Amadori and G. Marino. 2000. Effect of long-term administration of an immunostimulant diet on innate immunity in sea bass (Dicentrarchus labrax). J. Vet. Med. B Infect. Dis. Vet. Public Health 47: 745–751.

Beinfeld, M.C. 1995. Cholecystokinins/gastrin. *In:* F.E. Bloom and D. Kupfer (eds.). Psychopharmacology: the Fourth Generation of Progress. New York: Raven Press 585–594.

Bogwald, J., R.A. Dalmo, R.M. Leifson, E. Stenberg and A. Gildberg. 1996. The stimulatory effect of a fish hydrolyaste from atlantic cod Gadus morhua on atlantic salmon, Salmo salar, head kidney leucocytes. Fish Shellfish Immunol. 6: 3–16.

Bougatef, A., R. Ravallec, N. Nedjar-Arroume, A. Barkia, D. Guillochon and M. Nasri. 2010. Evidence of *in vivo* satietogen effect and control of food intake of smooth hound (Mustelus mustelus) muscle protein hydrolysate in rats. J. Functional Foods 2(1): 10–16.

Bourseau, P., L. Vandanjon, P. Jaouen, M. Chaplain-Derouiniot, A. Massé, F. Guérard, A. Chabeaud, M. Fouchereau-Péron, Y. Le Gal, R. Ravallec-Plé, J.P. Bergé, L. Picot, J.M. Piot, I. Batista, G. Thorkelsson, C. Delannoy, G. Jakobsen and I. Johansson. 2009. Fractionation of fish protein hydrolysates by ultrafiltration and nanofiltration: Impact on peptidic populations. Desalination 244: 303–320.

Cahu, C., I. Ronnestad, V. Grangier and J.L. Zambonino Infante. 2004. Expression and activities of pancreatic enzymes in developing sea bass larvae (*Dicentrarchus labrax*) in relation to intact and hydrolyzed dietary protein; involvement of cholecystokinin. Aquaculture 238: 295–308.

Cancre, I., R. Ravallec, A. Van Wormhoudt, E. Stenberg, A. Gildberg and Y. Le Gal. 1999. Secretagogues and growth factors in fish and crustacean protein hydrolysates. Marine Biotech. 1: 4489–494.

Chakraborty, P.K. and P. Madhavan. 1977. A pilot scale set up for the manufacture of fish hydolysate. Fish Technol. 14: 158–159.

Cordier-Bussat, M., C. Bernard, S. Haouche, C. Roche, J. Abello, J.A. Chayvialle and J.C. Cuber. 1997. Peptones stimulate cholecystokinin secretion and gene transcription in the intestinal cell line STC-1. Endocrinology 138 : 1137–44.

Cudennec, B., R. Ravallec-Plé, E. Courois and M. Fouchereau-Peron. 2008. Peptides from fish and crustacean by-products hydrolysates stimulate cholecystokinin release in STC-1 cells. Food Chem. 111(4): 970–975.

Cudennec, B., M. Fouchereau-Peron, F. Ferry, E. Duclos and R. Ravallec. 2012a. *In vitro* and *in vivo* evidence for a satiating effect of fish protein hydrolysate obtained from blue whiting (Micromesistius poutassou) muscle. J. Functional Foods 4(1): 271–277.

Cudennec, B., T. Caradec, L. Catiau and R. Ravallec. 2012b. Upgrading of Sea By-products: Potential Nutraceutical Applications. Elsevier (ed.) *In:* Advances in Food and Nutrition Research, Chapter 31, 65: 479–494.

Duarte, J., G. Vinderola, B. Ritz, G. Perdigon and C. Matar. 2006. Immunimodulating capacity of commercial fish protein hydrolysate for diet supplementation. Immunobiol. 211: 341–350.

Durand, P. and Y. Lagoin. 1983. Valorisation des sous-produits de la pêche. Bull. Inst. Pêches Marit. 330: 5–19.

Gauthier, S.F., Y. Pouliot and D. Saint-Sauveur. 2006. Immunomodulatory peptides obtained by the enzymatic hydrolysis of whey proteins. Int. Dairy J. 16: 1315–1323.

Gibbs, J., R.C. Young and G.P. Smith. 1973. Cholecystokinin elicits satiety in rats with open gastric fistulas. Nature 245: 323–325.

Gildberg, A., J. Bogwald, A. Johanssen and E. Stenberg. 1996. Isolation of acid peptide fractions from a fish protein hydrolysate with strong stimulatory effect on atlantic salmon (Salmo salar) head kidney leucocytes. Comp. Biochem. Physiol. 114B: 97–101.

Guerard, F. and R. Ravallec-Ple. 2001. Enzymic solubilisation of proteins from tropical tuna using alcalase and some biological properties of the hydrolysates. In "Focus in Biotechnology" (P. Thonart, M. Hofman), Vol. 39–50. Kluwer academic publishers.

Hou, H., Y. Fan, B. Li, C. Xue, G. Yu, Z. Zhang and X. Zhao. 2012. Purification and identification of immunomodulating peptides from enzymatic hydrolysates of Alaska pollock frame. Food Chem. 134: 821–828.

Hsueh, H.W. and M. Moskowitz. 1973. A growth factor for animal cells derived from peptone. Exp. Cell Res. 77: 376–382.

Igarashi, K., M. Hasama, S. Kimura and S. Enomoto. 2005. Effect of carnosine on the gastric secretion in rats. Trace Nutr. Res. 22: 121–124.

Kihara, M., M. Igarshi, T. Suzuki, F. Itou, S. Kozeni, M. Toyomane, J. Nakano and I. Yamai. 2011. Simulative effect of skipjack tuna soluble extract on pepsin-like protease in the stomach of rockfish (Sebastes schlegelii) using *in vitro* perfusion method. Comp. Biochem. Physiol. A 158: 444–449.

Kim, S.K., Y.R. Choi, P.J. Park, J.H. Choi and S.H. Moon. 2000. Screening of biofunctional peptides from cod processing wastes. J. Kor. Soc. Agric. Chem. Biotechnol. 43: 225–227.

Kotzamanis, Y.P., E. Gisbert, F.J. Gatesoupe, J. Zambonino Infante and C. Cahu. 2007. Effects of different dietary levels of fish protein hydrolysates on growth, digestives enzymes, gut microbia, and resistance to vibrio anguillarum in European sea bass (Dicentrarchus labrax) larvae. Comp. Biochem. Physiol A 147: 205–214.

Liddle, R.A. 1997. Cholecystokinin cells. Annual Review of Physiology 59: 221–42.

Mackie, I.M. 1982. General review of fish protein hydrolysates. Anim. Feed Sci. Technol. 7: 113–124.

Martínez-Alvarez, O., R. Ravallec, B. Cudennec, L. Guimas, C. Delannoy and M. Fouchereau-Peron. 2012. Effect of daily gavage with a collagen hydrolysate containing calcitonin gene-related peptide (CGRP)-like molecules on plasma CGRP-levels in rats. J. Functional Foods, In Press.

Merritt, J.H. 1982. Assessment of the production costs of fish protein hydrolysates. Anim. Feed Sci. Technol. 7: 147–151.

Nguyen, H.T.N., R. Pérez-Gálvez and J.P. Bergé. 2012. Effect of diets containing tuna head hydrolysates on the survival and growth of shrimp Penaeus vannamei. Aquaculture 324–325: 127–134.

Perdigon, G., M.E. Nader de Macias, S. Alvarez, G. Oliver and A. Pesce de Ruiz Holgado. 1986. Effect of perorally administered lactobacilli on macrophage activation in mice. Infect. Immun. 53: 404–410.

Picot, L., R. Ravallec, M. Fouchereau-Péron, L. Vandanjon, P. Jaouen, M. Chaplain-Derouiniot, F. Guérard, A. Chabeaud, Y. Legal, O. Martinez Alvarez, J.P. Bergé, J.M. Piot, I. Batista, C. Pires, G. Thorkelsson, C. Delannoy, G. Jakobsen, I. Johansson and P. Bourseau. 2010. Impact of ultrafiltration and nanofiltration of an industrial fish protein hydrolysate on its bioactive properties. J. Sci. Food Agric. 90: 1819–1826.

Quaglia, G.B. and E.J. Orban. 1987. Enzymatic solubilisation of protein sardines (Sardina pilchardus) by commercial proteases. J. Sci. Food Agric. 38: 263–269.

Ravallec-Plé, R., C. Charlot, C. Pires, V. Braga, I. Batista, A. Van Wormhoudt, Y. Le Gal and M. Fouchereau-Peron. 2001. The presence of bioactive peptides in hydrolysates prepared from processing waste of sardine (Sardina pilchardus). J. Sci. Food Agric. 81: 1120–1125.

Ravallec-Plé, R., L. Gilmartin, A. Van Wormhoudt and Y. Le Gal. 2000. Influence of the hydrolysis process on the biological activities of protein hydrolysates from cod (Gadus morhua) muscle. J. Sci. Food Agric. 80: 2176–2180.

Ravallec-Ple, R. and A. Van Wormhoudt. 2003. Secretagogue activities in cod (Gadus morhua) and shrimp (Penaeus aztecus) extracts and alcalase hydrolysates determined in AR4-2J pancreatic tumour cells. Comp. Biochem. Physiol. 134: 669–79.

Schwartz, M.W., S.C. Woods, D. Porte, Jr., R.J. Seeley and D.G. Baskin. 2000. Central nervous system control of food intake. Nature 404: 661–71.

Strader, A. D. and S.C. Woods. 2005. Gastrointestinal hormones and food intake. Gastroenterology 128: 175–191.

Tidwell, J.H. and G.L. Allan. 2001. Fish as food: aquacultures's contribution. EMBO Rep 2: 958–963.

Venugopal, V. 1994. Production of fish protein hydrolystes by microorganisms. *In:* A.M. Martin (ed.). Fisheries Processing: Biotechnological applications. London: Chapman & Hall Press 223–243.

Yang, H., Y. Liu, L. Ma and B. Kong. 2009. Hydrolyzing conditions and immunocompetence of sheep bone protein enzymatic lysates. Agricultural Sci. in China 8: 1332–1338.

7

Lipids from Marine Sources

Rocío Morales, María del Mar Muñío, Raúl Pérez Gálvez, Antonio Guadix* and *Emilia M. Guadix*

ABSTRACT

By-products of fish industry can account for up to 50% depending on the industrial process. They contain a high level of lipids: mainly polyunsaturated fatty acids (PUFAs) and a considerable fraction of EPA and DHA. These compounds present benefits for health such as: prevention of diabetes, allergy and some types of cancer, as well as anti-thrombotic, anti-arrhythmic and anti-inflammatory effects. This fact makes fish by-products an interesting raw material for the manufacture of concentrates of PUFAs and lipid structured. Indeed, the valorization of these materials has received much attention for food, pharmaceutical and nutraceutical applications. This chapter reviews a set of techniques currently available to concentrate crude fish oils (salmon, tuna, cod, sardine...) in PUFAs, such as urea complexation, supercritical extraction, enzymatic processes or low temperature crystallization. Besides this, this chapter presents structured lipids as a new approach to increase the therapeutic or nutritive value of the fish lipids recovered.

Department of Chemical Engineering, University of Granada, Severo Ochoa, s/n, 18071 Granada, España.
Emails: rocio_morales@ugr.es; mmunio@ugr.es; rperezga@ugr.es; eguadix@ugr.es; aguadix}@ugr.es
* Corresponding author

FISHERIES WASTE REUTILIZATION: AN ECONOMIC AND ENVIRONMENTAL CHALLENGE TO FACE

The world production of aquaculture has remained virtually constant during the last decade at around 130 million tons. However, with the rise of the world's population, the demand for fish resources intended for human consumption has increased by around 10 million ton (FAO 2012).

This imbalance can be alleviated, to some degree, by increase in aquaculture production, which currently provides 25% of the world's total production. This figure is expected to increase up to 50% by 2030 (Tidwell and Allan 2001). However, there is evidently a need for developing more efficient technologies able to supply human dietary requirements for lipids, proteins, polysaccharides and minerals, which are present to a large extent in aquatic resources.

Nowadays, the recovery yields of mechanical processing are not higher than 30–40% of the entire weight of the fish (Torres et al. 2007). In the European Union, 3.17 million ton per year of by-products are generated by fish filleting, salting and smoking and 1.5 million ton per year more comes from fish canning (AWARENET 2004). Presumably, around 50% of the total fish capture ends up as waste or by-products. Further, fish farming has a detrimental effect on the marine environment and fish waste management involves a great ecological impact cost, mainly due to the high perishability of fish materials. In addition, driven by the current hygiene standards, industries generate a significant amount of waste and wastewater (Arvanitoyannis and Kassaveti 2008).

Although it is quite common to grind and discard by-products along with by-catches (Kristinsson and Rasco 2000), a considerably percentage of them are used as raw material for (1) plant fertilizers, (2) livestock feeds and (3) added-value and specialty foods for humans. This kind of re-utilization provides a unique opportunity to fish processing industry to reduce disposal costs (http://www.earthprint.com/unep/download/2481.pdf.).

An approach to raise recovery yields involves a better usage of the by-products. Since they are a great source of protein, lipids and several nutrients, their use as fertilizers or simply as waste implies a great loss of compounds which are difficult to obtain and which have a high commercial or nutraceutical importance. For instance, there are formulations or even pharmaceutical applications based on such inexpensive marine feedstocks as the head, viscera, skin, tail or blood (Kim and Mendis 2006). Moreover, when the by-products are re-utilized for human food, the value of the product is multiplied five times (Gildberg et al. 2002). Hence, one of the most critical and important steps in order to achieve a better usage implies the development of new and economical technologies which allow the extraction and re-utilization of the protein or lipid content of the by-products. Also, the attempt to develop new technologies for recovery and purification will reap benefits related to both sustainability and human health (Ferraro et al. 2010).

The main target of this chapter is to describe the different lipids which are contained in fish as well as the lipidic profile of some species such as salmon and tuna. Also, the health benefits related to omega-3 consumption will be discussed. Finally, two different ways of marine lipids revalorization: omega-3 fatty acids concentrate and lipid structured will be explained.

MARINE LIPIDS

Among all compounds which are contained in fish, lipids are the ones which present a wider variety. Regarding their lipid content, fish species can be classified into three different groups: lean, semi-fatty and fatty. Lean fishes store their lipid content mainly in the liver, the cod being one of the most representative species. On the other hand, the fatty species such as mackerel, present a higher amount of lipid which is distributed throughout the body, under the form of subcutaneous and visceral adipose tissue (Pearson et al. 1977). As far as semi-fatty species are concerned, their lipid content varies between 2 and 10% and it is stocked in both organs and as subcutaneous deposits. Some authors have reported seasonal changes in both the lipid content and the fatty acid composition for fatty fish species such as cod and haddock (Falch et al. 2006).

The most common marine lipids can be divided into two main groups: triglycerides (also named as triacylglycerides) which are built by fatty acids and phospholipids. Additionally, there are other smaller groups as wax esters (which are basically storing fats) and fat soluble vitamins.

Fatty Acids and Triglycerides

As triglycerides are composed of fatty acids, the description of this kind of lipids is the most logical starting point. Fatty acids are carboxylic acids with long aliphatic tails which can be classified regarding their chain length, degree of unsaturation and the position of their double bonds.

When focusing on the chain length, fatty acids can be categorized as short (SCFA) when their aliphatic tail has less than six carbons; medium (MCFA) if the number of carbons ranges between 6 and 12 carbons and long-chain fatty acids (LCFA) with a tail longer than 12 carbons. The latter is the most common in fishes.

With regard to their degree of unsaturation, fatty acids are divided in two groups: saturated (with no double boundaries) and unsaturated. Furthermore, there are poly-unsaturated fatty acids (PUFAs) which contain more than two double bonds in their structure which are conjugated and separated by a methylene group. Finally, depending on the position of the first double bond, fatty acids can be classified as omega 3, omega 6 and omega 9 if the first double bond is 3, 6 and 9 carbon atoms respectively from the methyl end.

Taking advantage of this common structure of PUFAs, a shorthand nomenclature was developed. It is composed of the number of carbons of the chain, a colon, the number of double bounds, the Greek symbol "ω" or the letter "n" and finally the position of the double bond which is nearest of the methyl side of the molecule. By this terminology, eicosapentaeonoic acid (EPA) which is represented in Fig. 1a, b can be written as 20:5n3.

The most common PUFAs which are contained in fish are arachidonic acid (AA, 20: 4n-6) and linoleic acid (LA, 18: 2n-6), which is the AA metabolic precursor. In the case of omega 3 PUFAs, eicosapentaeonic acid (EPA, 20: 5n-3), docosahexaenoic acid (DHA, 22:6n-3) and linolenic acid (LA, 18:2n-6) are the main ones. Regarding saturated fatty acids (SFA), myristic (C14), palmitic (C16) and stearic acid (C18) are the most abundant in fishes.

Triglycerides (Fig. 1c) consist of a molecule of glycerol which is esterified to the sn-1, sn-2 and sn-3 positions by three fatty acids. Generally, monounsaturated and saturated fatty acids are located in the sn-1 and sn-3 positions of glycerol while PUFAs are situated in the sn-2 one. Nevertheless, there are a wide group of triglycerides which do not follow this rule, for instance, tridocosahexaenoylglycerol (tri-22:6n-3), which is the main component in the eye lipids of some fishes.

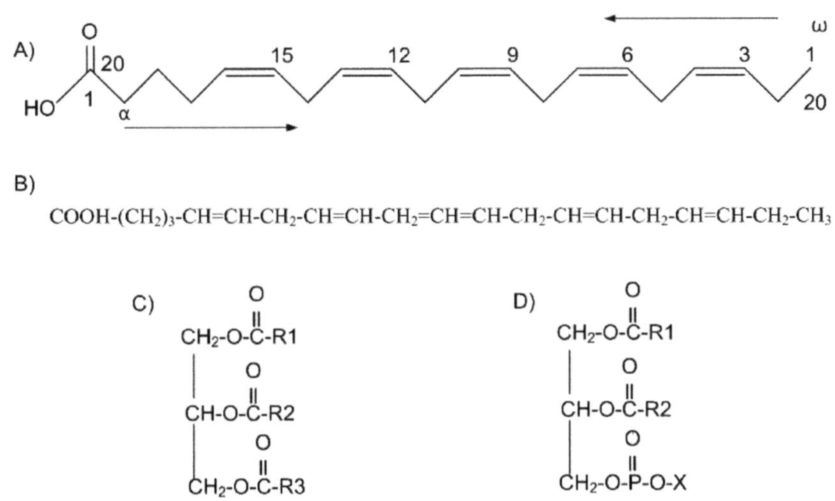

Fig. 1 Schematic representation of marine lipids.A) and B) EPA. C) Generic triglyceride and D) Generic phospholipid.

Phospholipids

Phospholipids are the main component of cell membranes. Most of them consist of a phosphate group, a diglyceride and an organic molecule (X) (Fig. 1d); their structure has a hydrophobic tail as well as a hydrophilic head.

In the case of phospholipids found in fish, monounsaturated and saturated fatty acids are usually esterified at the sn-1 position whereas the PUFAs are esterified at position sn-2. As with the triglycerides, there are several exceptions such as didocosahexaenoylphosphoglycerides which are found in the eye epithelia of fish.

Sources of Marine Lipids

This section deals with recent data related to the lipid content and PUFA composition of five species: tuna, mackerel, sardine, cod, and salmon. Further, as the main part of the by-products is the skin, head and viscera, a brief description of the lipid composition of these parts will be given. Taking into account that the lipid profile varies depending upon the season, some data related to this influence will be briefly described. These seasonal fluctuations are very relevant because they can cause several problems during the refinement and standardization of the oils. Finally, the lipid amount which can be obtained from the bones of different species is presented.

Tuna

Tuna's migratory behavior and its feeding habits (Grigorakis et al. 2002) are the two main reasons which explain its high content of DHA (C:22:6n3), which accounts for more than 20% of the total fatty acids. They mainly feed on small pelagic fishes and some crustaceans which have high levels of PUFAs. In addition, although DHA content in the stomach varied with the diet, the DHA levels in muscles remained high and constant. This fact could be related to their migration cycles (Saito et al. 2005).

As can be seen from Table 1, the total lipid amount varies widely during the year (35–50%). The maximum levels appeared from July to October. It is remarkable that the head is one of the organs with the highest count, even more than the muscle; furthermore its value is almost constant during all the seasons.

In Table 2 the lipid profile of the different organs in the month of January is shown (Salah and Saloua 2010). PUFAs percentage was the highest in winter whereas in autumn its value was minimal. In the case of SFA the evolution was the opposite. In all the organs the main fatty acid is

Table 1 Total lipid content (% of fresh weight) in the different organs of *Euthynnus alletteratus*. Modification of Salah and Saloua's data (2010).

Part	October	January	April	July
Liver	27.28	9.83	12.69	6.22
Red Muscle	12.2	6.46	5.71	4.38
Whitemuscle	8.95	3.72	2.97	1.7
Viscera	12.15	3.57	5.73	7.37
Head	14.14	14.26	11	11.75
Gonad	1.56	1.49	2.49	2.69

Table 2 Fatty acid content of tuna in January in different organs. Adapted from Salah and Saloua (2008).

Fatty Acid	Liver	Red Muscle	White Muscle	Viscera	Head	Gonad
14:0	0.87	1.31	1.88	2.63	3.71	1.35
16:0	20.82	17.41	17.96	19.19	13.69	21.10
18:0	8.87	10.06	5.70	6.94	2.40	8.15
16:1	1.35	1.30	2.90	1.66	7.54	2.09
18:1 n9	7.60	6.15	10.67	6.47	12.69	9.04
18:2 n6	1.20	1.00	1.40	1.51	1.69	1.38
20:5 n3	12.40	9.57	9.08	7.65	10.21	11.09
22:6 n3	31.19	40.14	31.18	23.57	34.02	28.74
Σ SFA	31.88	28.20	28.14	35.46	21.40	31.96
Σ MUFA	11.05	9.39	15.48	10.02	22.23	13.53
Σ PUFA	45.55	53.35	43.07	34.56	47.71	41.95

DHA followed by C16:0 and EPA. Due to this kind of lipid composition, tuna by-products are a suitable raw material for the obtaining nutraceutic products.

Salmon

The processing of red salmon (*Oncorhynchus nerka*) and the pink ones (*O. gorbuscha*) produces around 50, 000 ton of salmon heads and 25, 000 ton of viscera as by-products per year in Alaska (Oliveira and Bechtel 2005).

All parts of the salmon are rich in PUFAs, especially in omega-3 fatty acids (25–36%) (Oliveira and Bechtel 2005). A comparison of the fatty acid composition of the whole salmon, head and viscera is shown in Table 3. The percentage of SFA does not vary significantly as compared to the by-product with the whole fish, the main one being the palmitic acid (C16:0) of this fraction. Regarding PUFAs fraction, there is a significant variation

Table 3 Pink salmon fatty acid percentage of total oil extracted. Adaptation from Oliveira and Bechtel (2005).

Fattyacid	Wholefish	Head	Viscera
14:0	3.6	4.12	3.44
16:0	12.04	12.39	12.80
18:0	3.25	2.87	4.45
16:1	3.66	3.96	3.82
18:1n9	12.74	11.76	12.17
18:2 n6	1.73	1.67	1.38
20:4 n6	0.75	0.56	2.59
20:5 n3	7.70	7.56	10.93
22:6 n3	15.78	11.77	17.32
Σ SFA	16.51	17.47	16.59
Σ MUFA	40.05	42.69	34.19
Σ PUFA	38.50	33.26	44.46
Σ Omega-3	32.31	27.84	35.15

(33–44%) among all the parts. In all cases, DHA is the predominant fatty acid, varying from 11.8% in the case of head to 17.3% in the viscera. When focusing on eicosapentaenoic acid (EPA, C20:5 n3) it ranged from 7.6% in heads to 10.9% in the viscera.

Sardine

The lipid content of different parts of the Indian sardine (*Sardinella longiceps)*—head, muscle as well as by-products (bones, viscera and fins) were compared (Table 4).The total lipid percentage (wet weight basis) varied from 11–15%, being the fraction of by-products the one with the higher value. When focusing on the total percentage of PUFAs, the amount does not vary significantly; however, by comparing the EPA and DHA percentage in the case of waste, it is almost half of that in the head and flesh.

Although PUFAs amount is not as high as other species as tuna, sardine is a fatty fish with an adequate lipid profile for human consumption. Furthermore, the composition of sardine by-products makes it a good choice in order to process its by-products.

Mackerel

The case of mackerel is quite similar to that of the tuna; due to their migratory condition and feeding habits (crustaceans, mollusk and small fish) its level of DHA is quite high in the muscle.

Table 4 Percentage composition (wet weight basis) of different parts of Indian sardine (*Sardinella longiceps*). Adapted from Rai et al. (2011).

Fatty Acid	Head	Muscle	By-Products
14:0	10.8	10.6	9.2
16:0	25.5	25.2	35.3
18:0	5.3	4.9	9.5
16:1	11.2	10.2	8.9
18:1 n9	6.8	7.4	10.9
18:2 n6	1.2	1.2	1.0
20:5 n3	13.4	15.0	5.0
22:6 n3	8.2	9.2	5.6
Σ SFA	44.1	42.7	57.4
Σ MUFA	21.1	20.5	23.1
Σ PUFA	26.0	28.0	28.0
Total Lipid	13.05	10.97	15.10

Rai (2011) conducted a research about the lipid composition of the Indian mackerel in the muscle, head and by-products (bones, viscera and fins). The highest lipid amount was found in the head 13.60%, followed by muscle (8.79%) and finally by-products (6.20%). As is shown in Table 5, in all the cases the saturated fraction is higher than the unsaturated one. Furthermore, PUFA amounts are higher in the muscle followed by

Table 5 Fatty acid composition of the muscle, head and by-products of the Indian mackerel (*Rastrelliger kanagurta*). Adapted from Rai et al. (2011).

Fatty Acid	Muscle	Head	By-Products
14:0	6.5	6.6	8.1
16:0	25.0	33.4	34.3
18:0	7.9	8.7	9.6
16:1	9.0	8.8	9.4
18:1	10.6	10.0	10.
18:2 n6	1.2	0.8	1.1
20:4 n6	1.4	0.5	0.7
20:5 n3	8.6	4.2	4.7
22:5 n3	1.8	0.6	0.8
22:6 n3	15.3	7.9	7.5
Σ SFA	41.6	50.6	54.4
Σ MUFA	22.1	21.5	22.4
Σ PUFA	29	14.8	16.2

by-products and head. The predominant PUFA was in all cases DHA (22:6 n3), whose content in the muscle fraction was double as compared to other parts. EPA (20:5 n3).

Table 6 shows the evolution during the year of the lipid amount of mackerel fillets. The highest content was found in August while the minimum was observed in February. The saturated fraction ranged between 26.9% and 30.5% of total fatty acids, palmitic acid being the (16:0) the predominant one. Regarding the unsaturated fraction, its highest percentage was observed in July (43.4%). As has been previously mentioned, DHA (22:6 n3) was the major fatty acid of this group.

Table 6 Evolution of fatty acid profile. Total lipid is expressed as (mg/100 g edible fish) of total lipids of horse mackerel whereas the specific fractions were expressed as a percentage. [Modification of Bandarra et al. (2001)].

Fatty Acid	Spring	Summer	Autumn	Winter
14:0	58.4	102.4	169.6	135.8
16:0	294.5	768.6	966.3	647.6
18:0	101.4	255.2	291.6	204.7
16:1	96.2	235.4	302.5	190.3
18:1	290.5	902.6	1099.4	657.4
18:2 n6	31.0	41.3	56.4	38.8
20:4 n6	19.2	49.2	56.5	35.3
20:5 n3	10.1	28.2	38.5	25.5
22:5 n3	133.5	394.6	570.7	329.6
22:6 n3	318.0	759.4	858.5	598.3
Σ SFA	489.1	1195.5	1526.3	1040.7
Σ MUFA	490.0	1264.1	1610.3	1065.4
Σ PUFA	618.7	1570.5	1947.6	1250.4
Lipid Total	1597.7	4030.1	5084.2	3082.1

Cod

The most well known of all fish oils is cod liver oil. Nowadays, the main parts of the cod which are dedicated to human consumption are the liver and roe. Similar to other species, the lipid content of cod varies seasonally, in the case of liver from 58.5 to 76.1%. It has been estimated that around 1500 kg of lipids per day can be extracted from the by-products (Falch et al. 2006).

In Table 7 is summarized, the composition of the oil extracted from the liver and from the by-products (viscera and trimmings). In all the cases, PUFAs were the main bulk, with a percentage of 35–41% for liver, around 41% for viscera and 37–47% for trimmings. Although the viscera composition remains virtually constant throughout the year, liver and trimmings lipid content vary, with EPA and DHA amount the minimal in summers. These differences might be due to the spawning period.

Table 7 Levels of PUFA, MUFA, SFA, EPA and DHA (% of total fatty acids) and lipid content (% of wet weight) in the liver, viscera and trimmings of cod caught in Barents Sea. Adaptation from Falch et al. (2006).

	Liver			Viscera			Trimmings		
	Spring	Summer	Autumn	Spring	Summer	Autumn	Spring	Summer	Autumn
PUFA	41.0	35.5	36.0	41.3	41.2	41.7	36.8	38.2	47.3
MUFA	36.9	41.7	41.9	38.1	35.1	31.4	41.7	37.3	25.8
SFA	22.1	22.8	22.1	20.6	23.7	26.9	21.5	24.5	26.9
EPA	7.1	5.8	6.1	8.2	7.7	7.9	8.0	6.7	9.9
DHA	8.1	6.6	8.0	11.4	12.6	13.6	9.7	9.7	16.2

When comparing the main fatty acid composition of by-products and liver oil (Table 8), there are no significant differences. The energy value of by-products, calculated by Shahidi et al. (1997), presented an average value of 413.43 kJ/100g whereas the value estimated for the cod fillets was 344 kJ/100g. In the case of the cod, its by-products presented nearly the same quality as the edible parts.

Table 8 Fatty acid composition of by-products and cod liver. Adaptation from Shahidi et al. (1997).

Fatty Acids	By- Products	Liver
14:0	2.0	3.7
16:0	11.0	11.6
18:0	4.0	2.3
16:1	6.4	13.3
18:1 n9	20.3	21.7
18:2 n6	3.8	0.8
20:5 n3	8.9	7.7
22:6 n3	13.3	11.4

Lipid composition of bones

The chemical composition of bones varies significantly among the species, the lipid content varying the most: from 23 g/kg for cod to 509 g/kg for mackerel. Normally, fatty fish presents higher lipid levels in bones than lean fishes. Bones represent a significant part of the fish, around 10–15% of total fish biomass (Toppe et al. 2007). Though bones applications for human health have been widely studied as a protein and collagen source (Liaset et al. 2003, Kim and Mendis 2006); the lipid content in some species is not negligible at all. Nowadays several researches are conducted in order to extract the lipids from bones and use them as raw material for fishes diet (Toppe et al. 2007).

In Table 9, the different lipid compositions of several species are shown. Fatty species (trout, mackerel and salmon) have almost the double lipid amount than the lean one (cod). The low total fatty acid amount in the case of cod suggests that there is a significant bulk of phospholipids, whereas in the case of fatty fishes the situation is the opposite.

When comparing the lipid composition of bones to the lipid profile of the whole fish, very similar characteristics are observed. For instance, the most common fatty acids present in cod (Table 8) are C18:1n9 followed by DHA. This can also be observed in the case of salmon whose main fatty acids are C18:1 n9, C16:0 and DHA (Table 3) for whole fish and also for bones (Table 9).

Table 9 Fatty acids in fish bones in g/kg extracted lipid. Adapted from Toppe et al. (2007).

Fatty Acids	Cod	Salmon	Trout	Mackerel
14:0	9	44	51	63
16:0	87	119	130	157
18:0	28	23	23	28
16:1	12	54	53	42
18:1 n9	112	150	65	172
20:5 n3	46	65	41	52
22:6 n3	69	101	34	99
SFA	124	187	205	204
PUFAs	148	284	255	221
OMEGA-3	126	233	213	195
LIPID TOTAL	439	857	868	821

Health Benefits Associated with Marine Lipids

In human nutrition there are two essential fatty acids: **alpha-linolenic acid (ALA)** which is an omega-3 acid and **linoleic acid (LA)** which is an omega-6 fatty acid. These essential fatty acids cannot be synthesized by the human body but they play a very important role in the structure of cell membranes, and are precursors of eicosanoids, a group of compounds which are related to biological processes such as aggregation of blood platelets and cholesterol metabolism.

The positive health effects of ALA are because of three main factors (i) ALA is a precursor of EPA and DHA (Fig. 2), (ii) an ALA-rich diet increases the total omega-3 fatty acid of cell membrane which modify the behavior of the cell in a positive way (Nail et al. 1997, Tarpila et al. 2002) and finally, iii) it can dampen inflammatory processes by affecting the production of eicosanoids. These compounds can control inflammatory reactions and they

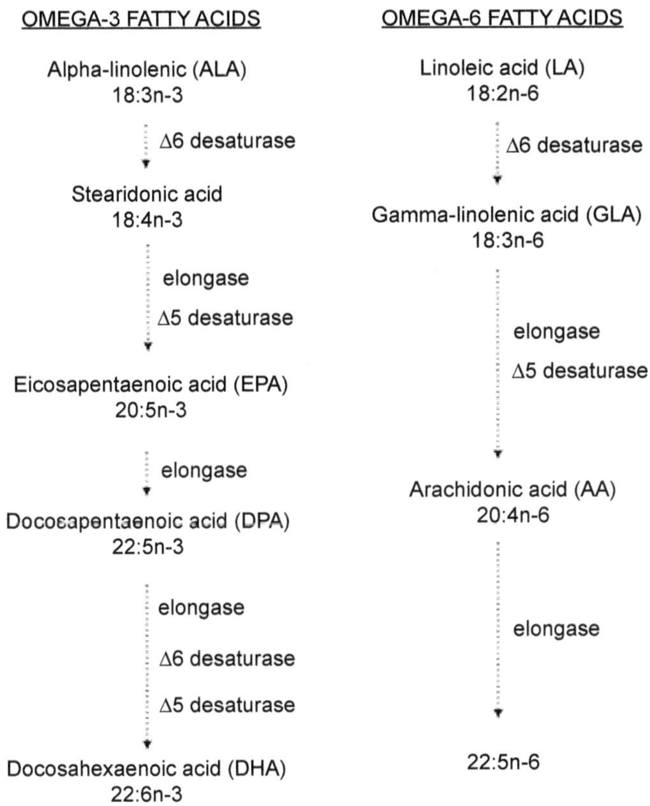

Fig. 2 Omega-3 and Omega 6 pathways [Adapted from Sahena et al. 2009].

are necessary in order to repair damaged tissue. Nevertheless, depending on the eicosanoids precursor they can either cause damage to the human body or be beneficial.

ALA has a double influence on eicosanoids: first, being a precursor of EPA, which is by itself an eicosanoid, it does not promote inflammation and decreases the risk of diseases when compared to omega-6 fatty acids. Second, as ALA is part of the cell membrane phospholipids it can block the metabolic pathway of omega-6 fatty acids (Magrum and Johnston 1983, Garg et al. 1990, Healy et al. 2000) which, as it will be explained later, has a detrimental effect on human health. However, ALA can minimize the production of cytokines, proteins which are liberated in response to infection or injury (Abbas et al. 1995) some of them being responsible for chronic diseases such as rheumatoid arthritis (Caughey et al. 1996).

As far as omega-fatty acids are concerned, the two main PUFAs obtained from LA are: gamma-linolenic acid (GLA) and arachidonic acid (AA). The first one is a precursor of some eicosanoids which have similar properties as those produced by ALA. However, the eicosanoids promoted by AA have negative properties such as promoting the aggregation of blood platelets and inflammatory reactions. As is shown in Fig. 2, the metabolic pathways of omega-3 and omega-6 depend on the same enzymes. Thus, an omega-6 rich diet unbalances the equilibrium between omega-3 and omega-6, fact which over-activates the immune system and causes several chronic diseases. Hence, it has been deeply studied how the dietary omega-3 PUFA ameliorates considerable biological and physiological functions of several systems such as the cardiovascular and the immune systems. PUFAs have demonstrated to play a key role against diseases as cancer or diabetes. Furthermore, it has also been proved that their positive effects are related to mental health, cognition and male fertility.

Cardiovascular benefits

Firstly, regarding cardiovascular and metabolic health, the consumption of EPA and DHA reduces the triglycerides levels as well as blood pressure. In addition, their ingestion increases the levels of cholesterol HDL and the stability of the plaques; as well as presenting anti-thrombotic, anti-arrhythmic and anti-inflammatory effects. Also, EPA and DHA prevent from several coronary heart diseases (Malik et al. 1996, Ruxton et al. 2005, Fedacko et al. 2007, Harris et al. 2008). Although the mechanisms by which PUFAs improve human health have not been totally elucidated, yet it is known that the anti-inflammatory properties are related to the fact that the synthesis of arachidonic acid is reduced (Rallidis et al. 2003). Some of these beneficial effects have been so clearly proved that several claims have been proposed to the European Food Safety Authority (EFSA). Those related to the maintenance of blood pressure and the triglyceride concentrations in normal levels have been approved; however, the claims which were associated with both, HDL and LDL cholesterol were declined by EFSA.

Omega-3 fatty acid's role in mental health

Regarding the effect of omega-3 PUFAs on mental health and cognition they have been found to exert a significant role on three different areas: influence in the early human development, in depression, as well as in the maintenance of the cognitive function in later life.

Early human development

In the course of the fetal development, the main effects related to EPA consumption are the improvement of both fluidity of cell membranes and functionality (Bourre et al. 1991). Furthermore, by the usage of DHA and arachidonic acid supplementation, the brain and retinal function can be enhanced (Ruxton et al. 2005). Further benefits related to hand-eye coordination, mental processing or a lower risk of wheezing have been previously enumerated (Helland et al. 2003, Dunstan and Prescott 2007, Miyake et al. 2009). However every health claim related to these benefits has been so far rejected by EFSA.

Depression

Several authors have described the benefits of omega-3 in neuropsychiatric disorders (Freeman et al. 2006, Parker et al. 2006, Schaefer et al. 2006). Although the mechanism is not yet known it is suspected that these fatty acids contribute to regulate the mood.

Maintenance of cognitive function: Alzheimer

Approximately 40 million people suffer from Alzheimer's disease where 65% are older than 60; it is expected that this figure will be doubled by 2030. Since this disease is the main cause of dementia, all the studies found in scientific literature are focused on studying the influence of omega 3 fatty acids in the prevention of Alzheimer's disease. It has been reported that the risk of suffering from Alzheimer disease can be decreased 60% by the weekly consumption of two fillets of any oil-rich fish (Morris et al. 2005).

In addition, a high content of omega 3 fatty acid in plasma may alleviate the decline in sensorimotor and complex speed (Dullemeijer et al. 2007).

Prevention of Type 2 diabetes

The studies which have been found in the literature show that preventing diabetes by means of omega-3 fatty acids intake is a controversial theme. Balk et al. (2004) published a review whose main conclusion was the lack of improvement of glucose tolerance when the consumption of EPA/DHA increased. By contrast, Nkondjock and Receveur (2003a) presented a study in which high intake of oily fish reduced the presence of type 2 diabetes, especially in people with obesity. This result is supported by Terry's work (2001), whose conclusion was that high oily fish and seafood intake caused a deep reduction in type 2 diabetes. Furthermore, positive effects in type 1 diabetes were found by Hu et al. (2003).

Nowadays, the majority of the studies have concluded that there is not enough evidence in order to prove any significant effect (Mori et al. 2000, Browning et al. 2007).

Prevention of Allergy and Asthma

The connection between a highly oily fish intake and a decrease in the risk of suffering from asthma or allergy was published by Schnappinger et al. (2009). What is more, the absence of omega 3 fatty acids during childhood was related to an enhancement of the risk of incurring asthma (Laerum et al. 2007).

Prevention of Cancer

Due to the wide variety of cancers, it is difficult to establish a direct link between the intake of omega 3 fatty acids and cancer prevention. Several reviews on breast, ovary, endometrial, stomach and pancreatic cancers have been so far published without obtaining enough evidence on the preventive activity of fish oil (Rose and Connolly 1999, Terry et al. 2001, Augustsson et al. 2003, Nkondjock et al. 2003b, Geelen et al. 2007).

Lipids Oxidation: A Setback to Deal with

The tendency of oxidation of PUFAs contained in the by-products is the main factor responsible for limiting revalorization. Oxidation can cause a decrease in the nutritional value as well as exerts a detrimental effect on the flavor and taste.

There are several factors which catalyze lipid oxidation such as heat, light, metals, enzymes or microorganisms. Furthermore, depending on how the oxidation process was generated there are different types: autoxidation, photooxidation, thermal or enzymatic oxidation; most of them dominated by radicalarial chemistry. The usual process of all which have been previously mentioned is autoxidation, a reaction among lipids and the atmospheric oxygen. In addition, there are enzymes (lipoxygenases) which can accelerate the oxidation process. The usual oxidation targets are unsaturated fatty acids which can appear as free fatty acids, acylclycerols, phospholipids or alkylesters (Shaidi and Zhong 2010).

In this chapter, we will describe the lipid autoxidation pathway as an example of how these mechanisms work. This process involves three stages: initiation, propagation and termination. Firstly, in initiation, an unsaturated lipid molecule loses a hydrogen atom due to the presence of initiators: heat, light, metal or metalloproteins (Frankel 2005). Because of this loss, the production of free radicals (R_1^*) begins (Fig. 3a). The free radicals react with oxygen producing peroxyl radicals whose main role is being the chain carriers of the reaction by reacting with others lipid molecules (Fig. 3b).

Depending on the nature of the element which causes the oxidation process, the pathways can vary. Examples of different mechanisms are shown in Fig. 4 for initiation caused by UV light or heat (Fig. 4a), when originated by oxidizing metals (Fig. 4b) and finally, when it is generated by reducing metals (Fig. 4c).

A) INITIATION

$$R_1H \xrightarrow{\text{initiator}} R_1{}^* + H^* \quad \text{and /or}$$

$$O_2 \xrightarrow{\text{initiator}} O_2{}^{*-} \longrightarrow HOO^* \xrightarrow{R_1H} R_1{}^* + H_2O_2$$

B) PROPAGATION

$$R_1{}^* \xrightarrow{O_2} R_1OO^* \xrightarrow{\text{a) } R_2H} R_2{}^* \xrightarrow{O_2} R_2OO^* \xrightarrow{\text{b) } R_3H} R_3{}^*$$

$$R_3{}^* \xrightarrow{O_2} R_3OO^* \xrightarrow{\text{c) } R_nH} R_n{}^* \xrightarrow{O_2} \ldots$$

C) TERMINATION

$$ROO^*, RO^*, R^* \longrightarrow \text{Non radical products (R-R, ROR. ROOR...)}$$

Fig. 3 Three steps of radical oxidation of oils: a) initiation, b) propagation and c) termination [Adapted from Shahidi et al. (2011)].

A) UV OR HEAT INITIATION:

$$R_1OOH \longrightarrow R_1O{}^* + OH^*;$$

$$R_1O{}^* \xrightarrow{R_mH} R_m{}^*; \quad OH^* \xrightarrow{R_xH} R_x{}^*$$

B) OXIDIZING METALS INITIATION:

$$R_2OOH \longrightarrow R_2OO^* + H^+; \quad R_2OO^* \xrightarrow{R_yH} R_y{}^*$$

C) REDUCING METALS INITIATION

$$R_3OOH \longrightarrow R_3O^* + OH^-; \quad R_3O^* \xrightarrow{R_zH} R_z{}^*$$

Fig. 4 Different initiation step regarding the nature of the source which origins the oxidation: (a) UV ligh or heat, (b) Oxidizing metals or (c) reducing metals. Adapted from Shahidi et al. (2011).

All these variations can occur at the same time. The propagation can consist of the repetition of these reactions several thousand times; this high number is achieved because of the creation of new radicals such as alkoxyl (RO*), peroxyl (ROO*), hydroxyl (*OH) and lipid radical (R*). Finally, the termination step (Fig. 3c) can be due to the absence of a hydrogen source or because of the interruption of the reaction by antioxidants.

A wide variety of products is obtained from lipid oxidation. For instance, hydroperoxides are the primary products of oxidation. However, as they are very unstable they continue reacting and producing secondary products such as aldehydes, ketones, alcohols, volatile organic acids, hydrocarbons or epoxy compounds. Normally, these secondary products are responsible for the undesirable odors as well as rancidity.

LIPIDS REVALORIZATION

As explained before, the utilization of the lipid content of fish by-products is motivated by their availability and low cost. Several processes have been developed aiming to extract fish oil from these materials such as urea complexation (Crexi et al. 2012) or supercritical fluid extraction (Perretti et al. 2007), among others. In this chapter, the main techniques to purify or up-grade the lipid fraction will be discussed with a special focus on PUFAs, which find many applications in functional foods and pharmaceutical preparations.

Also, marine lipids can be used as starting material to synthesize structured lipids which are classified as added value products (Muñío et al. 2008). Different ways to produce them will be exposed.

Production of Omega-3 Fatty Acids Concentrates

Concentrated fish oil is a type of fish oil which has been refined and modified aiming at increasing the omega-3 amount to at least 600 mg omega-3/g oil. Normally, omega-3 is concentrated as ethyl esters because by means of the usage of esters the processes are more economical and the yields higher (Nuria Rubio 2011).

Low temperature crystallization (winterization)

Basically, when considering winterization of edible oils, it consists of the separation of the oils into several fractions by means of their melting points. Oils which possess a higher melting point will form crystals which can be filtrated and separated from the liquid fraction. The main objective of the winterization of fish oils is to separate the saturated and unsaturated oils;

hence, oil with a high PUFA amount can be obtained. Organic solvents (hexane, acetone) are used in order to increase the velocity of the mass transfer as well as the temperature of the process (Alasalvar et al. 2011).

The melting point of fatty acids depends on several factors, the most important of them being the chain length and saturation degree. Saturated and monosaturated acids, have higher melting points than PUFAs. Thus, the former crystallize while PUFAs remain in the liquid phase.

Briefly, the main parameters of crystallization are: temperature, organic solvent, process time as well as the organic solvent:oil ratio. Wanasundara and Shahidi's (1999) results showed how the recovering yields of PUFAs were much higher by the usage of hexane rather than isopropanol. Besides, it has been proved that when the solvent:oil ratio are high the separation process is more effective (López-Martínez et al. 2004) in spite of the economic disadvantages. Furthermore, a previous saponification of TAGs in order to obtain FFAs increases the yields.

Urea complexation

Urea complexation, which is based on the formation of solid complexes of straight-chain organic compounds with urea, is one of the concentration methods with higher yields, achieving a concentration of 85% between EPA and DHA (Haraldsson 1983, Gámez-Meza et al. 2003, Liu et al. 2006).The complexes of urea are usually tetragonal-shaped, however, in the case of straight-chain aliphatic compounds the configuration varies, developing hexagonal prisms in whose inner channel several organic compounds (hydrocarbons, alcohols, esters and fatty acids) are fixed (Breivik 2007). In case of the addition of a mixture of fatty acids or esters to urea, the more saturated compounds are trapped more easily. Consequently, the solid complexes can be removed, leaving a liquid fraction enriched in polyunsaturated acid or esters.

Briefly, this method begins with an alkaline hydrolysis of the triglycerides aiming to obtain free fatty acids. Subsequently, the free fatty acids are added to an ethanolic solution of urea. The complexation process then takes place by cooling and, finally, crystals are separated by filtration (Hayes et al. 1998, Crexi et al. 2012). One of the most relevant variables to control is the molar ratio between alcohol and triglycerides during the hydrolysis. Although the stoichiometric is 3:1, the reversible behavior of the reaction makes necessary the use of an excess of alcohol (Meher et al. 2006).

Linko and Karinkanta (1970) carried out experiments with herring oil and reached the conclusion that the complexation process improved (i) when the length of the carbon chain increased, (ii) the number of double bonds decreased or (iii) the distance from the first double bond to the acid group increased. As a consequence, all PUFAs were not concentrated in the same proportion and there were small fractions of EPA and DHA which were lost in the complexes.

Urea complexation presents several advantages as compared to winterization. As complexes are much more stable than crystals which are produced in winterization, the filtration step can be carried out at higher temperatures, which involves a higher economic yield. In addition, organic solvents used for urea complexation are less toxic than other organic solvents required in other concentration method such as hexane (winterization), hence, it is an environmental-friendly process. Further, since the separation is based on chemical (configuration of the organic compounds) rather than on physical properties (melting point or solubility) this method is preferred by many researchers (Wanasundara and Shahidi 1999).

Distillation methods

This process, along with urea complexation, is the most employed to concentrate omega-3 from fish oil on large scale. The separation takes place due to the differences in the boiling point and molecular weight of the compounds at low pressure. Distillation consists of two steps: firstly, esterification of fatty acids and subsequently separation of ethyl esters at low temperature and vacuum (10^{-3}–1 mbar). It must be carried out at low temperatures, otherwise undesirable reactions such as hydrolysis, thermal oxidation, polymerization or isomerizationmay occur (Ackman et al. 1988, Wijesundara et al. 1989).

Due to the difficulties that high temperature causes, PUFAs are distillated by means of short-path distillation (molecular distillation) technique which is characterized by short exposure time of the compounds and the short distance (20 to 70 mm) from the evaporator to the condenser. The main drawback of this process is the high equipment (high vacuum pump, distillator, condenser…) cost. On the other hand, the production which can be achieved with short path distillation is 100-200 kg per square meter evaporator an hour (Martins et al. 2006).

For industrial production, there are two different operation conditions (centrifugal molecular distillation and thin film evaporator) depending on how the thin film of the mixture which is going to be evaporated is obtained. When centrifugal molecular distillations taking place, the film is produced by the usage of a disk which rotates at high speed. As the disk is heated, the more volatile compounds are evaporated. The selective fractionation is controlled by changing the rotation speed, which varies the film thickness and the residence time. The other configuration is a thin film evaporator which distributes the compounds by the use of a mechanical agitator. By the addition of wiping elements, the heat transfer can be notably improved. A schematic diagram of the same is shown in Fig. 5.

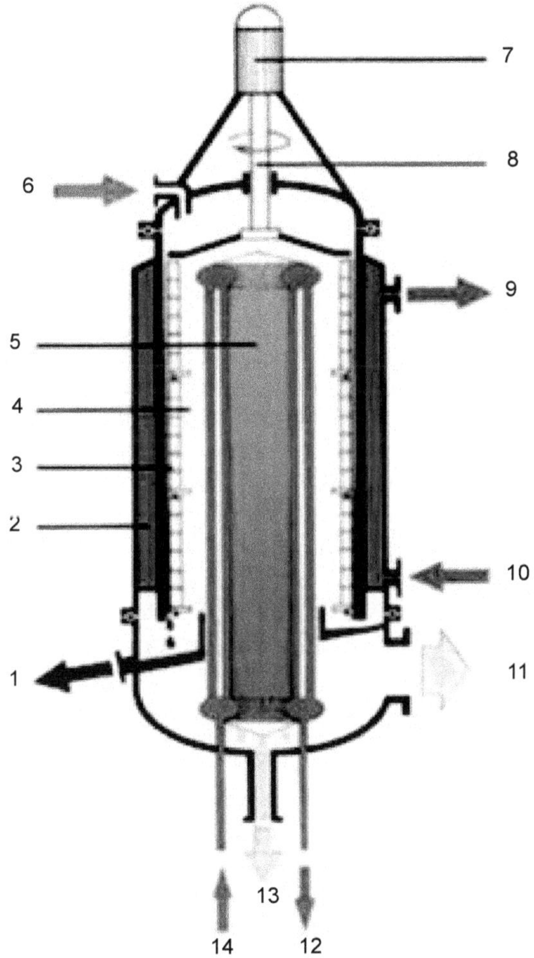

Fig. 5 Section of a short path distillator. 1. Residue nozzle. 2. Heating jacket. 3. Wiper system. 4. Vacuum chamber. 5. Condenser. 6. Feed nozzle. 7. Motor. 8. Shaft with distribution plate for feed. 9. Heating oil (out). 10. Heating oil (in). 11. Vacuum connection. 12. Cooling liquid (out). 13. Distillate nozzle. 14. Cooling liquid (in). Figure reproduced from UIC GmbH, Alzenau-Hörstein, Germany (http://www.dax-consult.dk/documents/fish-oil-industry.pdf).

By means of this process the EPA and DHA percentage of an initial sample with 25% can be enhanced to 50% (Skaliotis 2011). Hence, high purity concentrates are obtained in a two-step process or in combination with selective enzymatic transesterification (Breivik et al. 1997).

Supercritical (SC) extraction

In general, supercritical fluid extraction comprises two main steps: first, the extraction of the components of interest and second the separation of the compounds from the supercritical matrix. One of the most common supercritical solvents is carbon dioxide because of its properties such as non-reactivity, easy availability and low cost. Furthermore, when it is used as an extractor with PUFAs, it can reduce oxidation and decomposition of oils.

As fish oil is composed of a wide variety of compounds with different chain lengths and saturation points, the fractionation of a specific compound is virtually impossible. However, separating fatty acid methyl or ethyl esters from fish oil triglycerides is a feasible option which is able to increase the concentration of EPA from 14.5 to 48.2% (Eisenbach 1984).Moreover, when urea crystallization was conducted before supercritical fractionation the percentage was enhanced to 95% (Nilsson et al. 1989).

The most relevant variables which govern SC fractionation are press, temperature, solvent flow rate and process time. In addition, the study of the phase equilibrium of the mixtures is essential in order to design the fractionation process (Esquível et al. 1997, Dunforda et al. 1998). Solubility of the oil increases notably with pressure at a constant temperature. On the one hand, it causes an increase of the density of the solvent which produces a decrease in solubility, but, on the other hand, the vapor pressure of the solute increases, a fact which is responsible for the rise of the solubility (Brunner 1994).

Perretti et al. (2007) conducted the optimization of the main variables of fractionation in a column with three sections, with the coolest at the bottom. By modifying the temperature and pressure along the column obtaining the composition required in the market was obtained. In addition, fractionation of oil obtained from sardine waste has achieved purity up to 95% of fatty acids methyl esters and yields around 45% (Létisse and Comeau 2008). In this case, the extraction was carried out in four steps of 45 min each, at 60°C, with a flow rate of 1 mL/min and the density of CO_2 was increased from 500 to 800 kg/m^3. Furthermore, a mathematical model for the supercritical extraction with CO_2 has been developed and validated for fish oil ethyl esters mixtures (Martín and Cocero 2007).

Although supercritical extraction presents better properties and a more environmental-friendly process than the previously explained solvent or enzymatic concentration method, the main drawback, when scaling up, is the amount of CO_2 which is consumed (solubility of oils is around 10 g/kg) (Sahena et al. 2011). However, this quantity of CO_2 can be minimized by the usage of the pressure swing technique (Zaidul et al. 2007), a method which, by combining cycles of pressurization and depressurization steps, can achieve the same yields as a continuous supercritical fractionation.

Enzymatic processes

The production of omega-3 concentrates by means of enzymatic processes is basically based on the usage of specific enzymes (lipases) which are responsible for reactions such as hydrolysis or transesterification of triglycerides. The main difference between these processes is that, in case of hydrolysis, PUFAs obtained from triglycerides are in the form of free fatty acids whereas when transesterified they are under the form of esters (Haraldsson 1997, Lyberg and Adlercreutz 2008, Okada et al. 2008, Carvalho et al. 2009).

After the reaction, a separation process is usually required; being membrane filtration, molecular distillation or urea complexation the most used techniques in case of omega-3 concentrates production (Wanasundara and Shahidi 1998, Carvalho et al. 2003). As mentioned before, esters are more easily fractionated than free fatty acids; a fact which makes enzymatic esterification a more competitive and economic method.

The main difficulty of enzymatic processes is related to the fact that the selectivity of lipases can vary with the distribution of fatty acids in the triglyceride (sn1, sn2 or sn3 position) and even the glyceride structure (Tanaka et al. 1993, Mbatia et al. 2010). As a result, sometimes, a lipase is useful in concentrating EPA and DHA in some species and not effective in others. This is the case of *Aspergillys niger* which was able to increase DHA amount in cod liver (Hoshino et al. 1990) but did not have any significant effect with tuna (Tanaka et al. 1992) or menhaden oil (Wanasundara and Shahidi 1998).

As far as hydrolysis is concerned, several researches have been conducted in order to improve the purity and yield of omega-3 concentrates. For instance, a method in which sardine oil was concentrated by means of, first, hydrolysis with *Pseudomona cepacia*and, secondly, urea complexation achieved an omega-3 purity of 86.58% and a yield of 78% (Gámez-Meza et al. 2003). Furthermore, Kojima et al. (2006) developed a process able to concentrate DHA or EPA from a mixture of fish oil. The selectivity can vary by the choice of the enzyme: the usage of HU-lipases produces higher concentrates of EPA whereas the use of AK-lipase increased the DHA percentage. In both cases, hydrolysis was followed by urea complexation as a separation process. In addition, a final step in which fatty acids changed to the form of triglycerides must be conducted.

Enzymatic esterification also requires the reconversion of fatty acids esters to triglycerides or at least to a mixture of mono-, di- and triglycerides (TG*).

This reconversion can be carried out in two different ways: (1) with a two-step enzymatic reaction in which fatty acid esters are hydrolysed and converted into fatty acids and then esterified with glycerol or (2) via a one-step reaction in which a direct transesterification of fatty acid esters takes place.

Enrichment of oil extracted from the Nile perch viscera was conducted by both esterification of free fatty acids as well as by hydrolysis of ethyl esters of fatty acids, by using *Pseudomonas cepacia.* Lipases were able to discriminate between EPA and DHA with higher selectivity when they were ethyl esters rather than in FFA form varying the recovery yield of EPA from 79% to 92% (Mbatia et al. 2011). In both cases, the products were more enriched when compared to hydrolysis of natural fish oil.

Structured Lipids

Generally, structured lipids (SLs) are defined as triacylglycerides (TAGs) obtained by chemical or enzymatic process that present a certain fatty acid composition to achieve a determinate purpose, taking into account the different positions of glycerol.

Provided that till the moment those aims are directed to human nutrition, the most important aspects are the nature of the fatty acids involved and their position in glycerol. About position, two different positions can be distinguished according to intestinal absorption, extreme or primary positions (*sn*-1 and *sn*-3) and intermediate or secondary position (*sn*-2). In Fig. 1c glycerol structure can be clearly appreciated, where R, R′ and R″ are alquil radicals are alquil radicals. The condition for being homogeneous triacylglyecrides is that the same fatty acid is located at the three positions (R, R′ and R″ are the same fatty acid). However, when different fatty acids are located at the three positions (R, R′ and R″ are different fatty acids), they are named as heterogeneous triacilglycerides.

On the other hand, the afore-mentioned triacylglycerides can be transformed by chemical or enzymatic reactions into different species according to the position of the hydrolyzed fatty acid. Monoacylglycerides (MAG) are obtained by hydrolyzing two fatty acids from TG molecule, diacylgycerides (1, 2-DAGs or 1, 3-DAGs) result from hydrolysis of just one of the fatty acids and in both cases, hydrolyzed fatty acids remain as free fatty acids (FFAs) in the media.

Since SLs are focused at human nutrition, the raw material used in its production must be vegetal or animal natural TAGs. Moreover, the usual synthesis process of TAGs is enzymatic due to the delicate structure of fatty acids, especially polyunsaturated fatty acids (PUFAs).

Structured lipids began to be used in nutrition as medium chain triacylglycerides that were prepared from vegetable fatty acids: saturated fatty acids from caproic (6:0) to lauric (12:0) acids. Mentioned triacylglycerides are used to provide rapid energy food to patients with mal-absorption syndrome (Babayan 1987) or to produce hypocaloric food for dietetic purposes. Nevertheless, currently these kinds of TAGs are focused to be used as methods to provide functional fatty acids for treatment of disease or for certain deficiencies.

Therefore, structured lipids are currently defined as triacylglycerides containing short- or medium-chain fatty acids and long chain ones in the same glycerin (Akoh 1995). Even when they present in any distribution (random structured lipids), the preferred distribution is the one that has medium chain fatty acids at sn-1, 3 position and essential and functional fatty acids at sn-2 position of triacylgyceride (MLM structured) due to TAGs human metabolism.

In the last few years, enzymatic processes using lipases have firmly appeared to obtain specific lipids for each age and physiological state (Merolli 1997, Fitch Haumann 1997). But lipid's technological use is more complex that for other macronutrients, mainly because of the poor solubility of their constituents in water.

Structured lipids synthesis can be developed basically by two kinds of processes: chemical and enzymatic. Occasionally, chemical processes include some kind of enzymatic reaction.

Digestion and absorption of lipids

The importance of developing structured lipids is basically related to the fact that the efficiency of absorption depends not only on the composition but also on the positional distribution of fatty acids.

As shown in Fig. 6, lipid digestion begins in the mouth with the usage of lingual lipase, an enzyme which hydrolyses triacylglycerides (TAGs) to diacylglycerides (DAGs), monoacylglycerides (MAGs) and FFAs. This lipase shows a major specifity towards short and medium fatty acids as well as towards fatty acids in the sn-3 position; being the main products 1, 2 DAGs and FFAs (Small 1991, Christensen et al. 1995).

In the stomach, digestion continues with the gastric lipase, an enzyme which is even more specific towards short and medium fatty acids than lingual lipase. In this case, positions 1 and 3 of TAGs are hydrolyzed producing FFAs, MAGs and DAGs. As short and medium free fatty acids have a high solubility in aqueous media they are absorbed through the stomach and transported to the liver. Mainly, they act as a rapid source of energy (Borum 1992).

The compounds which are not soluble enough are finally hydrolyzed in the small intestine by means of the pancreatic lipase, a 1, 3-specific enzyme (Small 1991, Bernadier 1995). However, it presents a very low activity when reacting with sn-1 and sn-3 long PUFAs (Christensen et al. 1995). Then, fatty

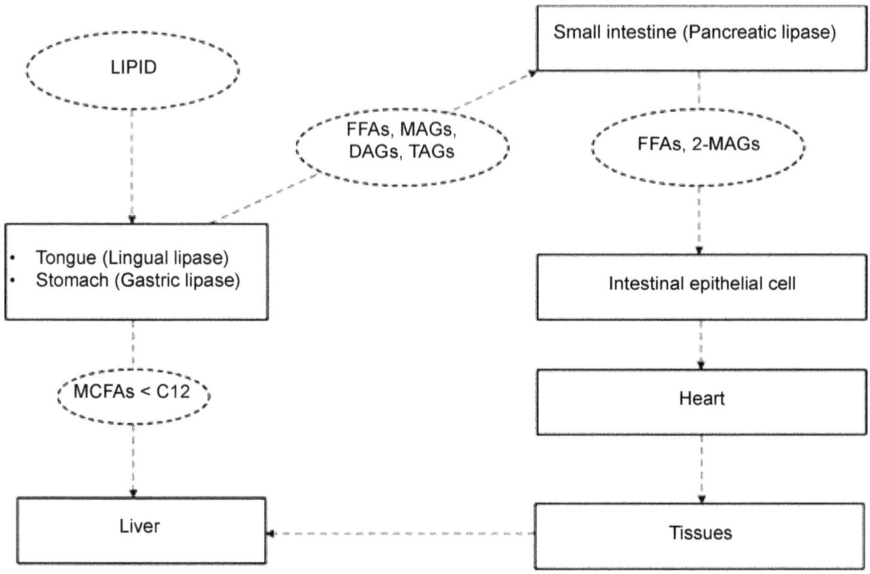

Fig. 6 Digestion and absorption of lipids [Adapted from Willis and Marangoni (1999b)].

acids and 2-MAGs formed micelles with bile salts and, subsequently, they are absorbed by the intestinal mucosa.

The position of long chain saturated fatty acids affects their solubility. For instance, palmitic acid is weakly absorbed because at body temperature it is solid and forms insoluble soaps in the intestine; however, as a 2-MAG, palmitic acid can be easily absorbed (Innis et al. 1995). By the placement of long chain PUFAs at the sn-2 position, it is obtained 2-MAGs which are poorly hydrolyzed by pancreatic lipase and which present a better absorption. However, in the case of long chain PUFAs in positions sn-1 and 3, a hydrolysis by carboxyl ester hydrolases is required (Small 1991).

After absorption from the intestines, new TAGs are produced by the reaction of FFAs and 2-MAGs. Consequently, they are combined with phospholipids and apolipoproteins in order to form chylomicrons and be transported through the lymphatic system to the general circulation.

Chemical synthesis

In order to obtain structured lipids by chemical synthesis, reactions are developed at very high temperature. The time used for the synthesis is very long and reactions are catalyzed by alkaline metals or alkaline metal alkylates (Akoh 1995). This kind of catalysis is cheaper and easily scalable but has no specificity and fatty acids distribution in the final product cannot be controlled (Willis and Marangoni 1999a).

Halldorsson et al. (2003) proposed a chemical-enzymatic method to obtain structured TAGs. The method has two steps: first, enzymatic esterification of glycerol with a medium chain fatty acid and Lipozyme lipase to obtain 1, 3-DAGs with 90% yield after purification. Later, chemical agents can be used to introduce in the molecule formed, pure EPA and DHA. With both reactions and after silica gel chromatography to remove pure fatty acids excess, final yields of around 90–95% can be achieved.

Lower yields and high purity can be obtained using just chemical synthesis of structured TAGs with AA or DHA and lauric or palmitic acids (Fauconnot 2005), reactions can be carried out in the presence of PUFAs, saturated MAGs and a couple of chemical reagents. Structured lipids as APA and DPD yields could range between 63 and 68% with purity higher than 98% in each case (P: palmitic acid, A: arachidonic acid and D: docosahexaenoic acid).

Enzymatic synthesis

Enzymatic processes present a great advantage, fatty acids distribution in the final product can be controlled because of lipases regiospecificity and specificity towards fatty acids.

Structured lipids can be synthetized by enzymatic catalysis in two ways:

- One-step processes: such as acidolyisis and interesterification.
- Two-step processes: like alcoholysis + esterification or hydrolysis + esterification.

Among the first kind of processes, acidolysis is the one most used while alcoholysis + esterification is the most commonly used among two step processes.

Acidolysis

Acidolysis reaction occurs between an ester and an acid with the substitution of acyl-ester by free fatty acids (Fig. 7a).

This reaction can be interpreted as a two-step process that requires a previous removal of a fatty acid from the TAG (hydrolysis) followed by an introduction of another fatty acid into the free position (esterification) (Ainsworth et al. 1996).

This kind of reaction has been used in EPA and DHA concentration from marine oils glycerides (Yamane et al. 1992, 1993) and in SLs production, type MLM (Hita 2007).

Structured lipids synthesis by acidolysis has been carried out by several authors during last few years. Due to the presence of just one step, it can

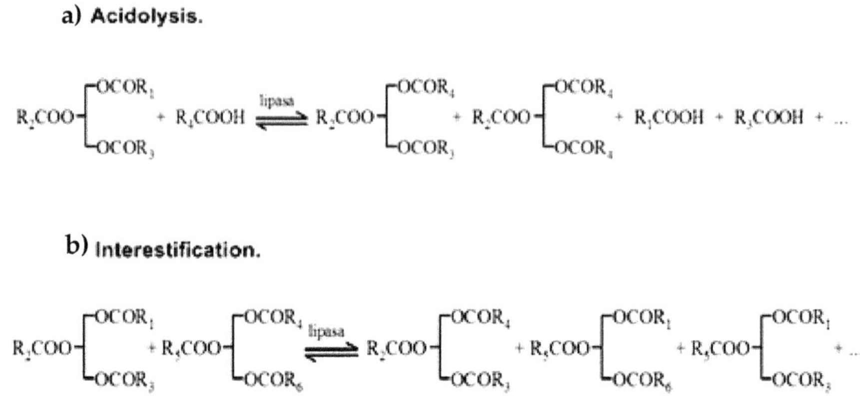

Fig. 7 Generic reactions of (a) Acidolysis and (b) Interestification.

be considered the easiest way to produce SLs. Yields around 65% were obtained by Shimada (1996a,b) when catalyzed acidolysis was developed with tuna oil rich in DHA and caprilic acid using *Rhizopus delemar* as the lipase. An important aspect of this process is that almost all DHA remained at the central position of TAG.

Xu et al. (1998) also synthetized structured lipids using fish oil, capric acid and Lipozyme IM® lipase in a batch reactor. After six hours of reaction, more than 60% incorporation of 10:0 was achieved but also 10% acilmigration from sn-1, 3 to the central position. The same authors (Xu et al. 2000a) obtained 40% incorporation of 8:0 in a packed bed reactor with just 3% of acilmigration. The same incorporation around 40% of oleic acid was achieved when canola oil was used (Xu et al. 2000b).

Immobilized lipase IM60®, was used as a biocatalyst for the incorporation of capric acid (C10:0) into menhaden fish oil concentrate (34.7 mol% EPA and 34.4 mol% DHA), by means of an acidolysis reaction. It was found that the presence of hexane was beneficial for the incorporation of capric acid, while it was detrimental to the concentration of EPA and DHA. Nevertheless, this drawback can be alleviated by increasing the capric acid/oil molar ratio (Jennings and Akoh 2001).

Structured lipids synthesis from cod liver oil, caprilic acid and Lipozyme IM® in different systems was studied by Camacho Páez et al. (2002). In the first system a batch reactor was used and temperature influence in the caprilic acid incorporation was analyzed. In the second system a packed bed reactor with recirculation was employed to study the kinetic process. Finally, the third system used was a packed bed reactor with no recirculation and MLM type TAGs was obtained with 57% of 8:0 and 44% of EPA and DHA at the central position.

A similar study was developed by González et al. (2004) using caprilic acid, a commercial oil rich in EPA and Lipozyme IM®. In function of operating mode used TAGs obtained had 51–64% of caprilic acid and 9.6–19.6% of EPA. When using the same conditions but with a packed bed reactor with recirculation, the structured lipids presented 59.5% caprilic acid, 9.6% EPA, 2.2% DHA and 11.8% oleic acid.

Acidolysis can also be carried out with tuna oil, caprilic acid and Rd lipase in a packed bed reactor with recirculation. In the absence of solvent structures TAGs with 51% caprilic acid and 13% DHA were obtained after 73 h. The structure analysis showed that 91% of caprilic acid was located at extreme positions and 51% of DHA at central position. In order to purify SLs, several extractions with water/ethanol mixture was used and TAG purity resulted to be 93–96% with a separation yield of 85% (Hita et al. 2007, 2009).

On the other hand, Camacho et al. (2007) proposed a kinetic model for acidolysis of different TAGs and caprilic acid in a packed bed reactor with Lipozyme IM®. The data obtained was adjusted very well by the proposed model just when working with TAG concentration lesser than 0.1 mol/L. At higher concentrations a significant decrease in reaction rate was observed.

Interestification

Interesterification reaction occurs between two esters with an exchange of acyl group. If ester was a TAG the reaction is exemplified in Fig. 7b. New TAGs that can exhibit different properties from the original TAG, are formed. For example, that is the way to produce n-3 PUFA-enriched TAGs using Lipozyme lipase (Haraldsson et al. 1989).

Several authors obtained MLM-structured lipids using cotton seed oil, tricaprilin, tracaprin and trilinolein (Soumanou et al. 1998b, Fomuso and Akoh 1998).

Later, Han et al. (1999) used tricaprilin and EPA ethyl esters with several lipases and checked that water activity had no effect on the composition of the final product when specific lipases are employed. Also, the first EPA incorporation was faster than the later ones because of steric impediments.

Interesterification between triEPA and ethyl caprilate can be carried out with Lipozyme™ at 40°C. Irimescu et al. (2000) reached a 91% molar of structured lipid (CEC, C: caprilic acid and E: EPA) in the final reaction mixture.

When using tripalmitin and saturated fatty acids (2 to 14 carbon atoms) ethyl esters in the presence of *Carica papaya* lipase act as catalyzer. Gandhi and Mukherjee (2001) observed that incorporation was higher when increasing ethyl esters chain length.

Two steps process: Alcoholysis and Esterification

Alcoholysis reaction happens between an ester and an alcohol to produce alkyl group of the ester substitution by the alcohol one. If ester was a TAG the generic reaction is shown in Fig. 8a.

This reaction can be interpreted as a two-step process that requires a previous removal of a fatty acid from the TAG (hydrolysis) followed by the esterification of the alcohol and the fatty acid removed (Ainsworth et al. 1996). When alcohol used is glycerine, the final product obtained is a mixture of MAG, DAG and TAG.

Because water is not produced in alcoholysis, planning for water removal is not necessary. Nonetheless, main alcoholysis problem is that an excess of alcohol is needed to achieve a high yield and that excess should be recovered later.

Esterification reaction happens between a fatty acid and an alcohol to produce an ester. If glycerin was used instead of alcohol, the reaction is shown in Fig. 8b. The typical reaction products are MAGs, DAGs, TAGs and non-esterified free fatty acids. Using lipases acyl-selectivity towards some fatty acids, esterification can be applied for the GLA and DHA concentration (Shimada et al. 1997a, 2001a,b). For DHA concentration lauric alcohol and free fatty acids from tuna oil can be used to achieve an increase of 43% in the DHA mixture content. The same reaction can also be employed to obtain enriched TAGs using glycerol and PUFAs (Esteban et al. 1998).

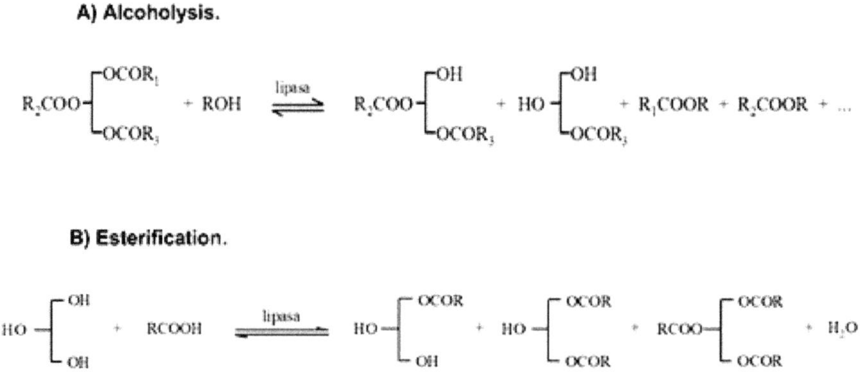

Fig. 8 Generic reactions of the two-steps process (a) Alcoholysis and (b) Esterification.

Several authors have carried out structured lipid synthesis using the two-step process (alcoholysis and esterification) and came to the conclusion that it has some advantages. Irimescu et al. (2001b) concluded that this process is faster and higher conversions can be achieved in the absence of toxic solvents.

Soumanou et al. (1998a) also reported that higher concentration of MLM structured lipids were obtained using the above-mentioned two-step process as compared with acidolysis.

Alcoholysis was developed using several homogeneous oils and natural oils with *Rhizopus delemar* immobilized in a support and with organic solvents. After purification, 2-MAGs were esterified with caprilic acid and *Rhizopus miehei* and the final product had more than 90% of caprilic acid at the extreme positions and more than 98% of PUFAs at the centre.

Irimescu et al. (2001a) used two homogeneous TAGs, triEPA and triDHA, to produce MLM structured lipids. *Candida antarctica* lipase was used in alcoholysis and after 2-MAGs esterification with *Rhizopus miehei*, TAGs were obtained with more than 85% yields. The whole process lasted less than 8 hours. No isomers were formed in the synthesis and yields slightly decreased when triEPA was used as a substrate. If heterogeneous oils (tuna oil) are used in the same process, after 3 hours of complete reaction, MLM structured lipids with purities higher than 85%, 45% of DHA content and a purification yield of 71% (Irimescu et al. 2001b) were obtained.

Torres et al. (2003a) synthetized structured TAGs from fish oil. After the alcoholysis step and esterification with *Pseudomonas cepacia*, TAGs with 55% of conjugated linoleic fatty acids and 18% of EPA and DHA were obtained.

The two-step process to obtain structured lipids was also developed using fish oil as the starting substrate and several lipases (*Candida antarctica*, *Rhizopus delemar* and lipase D) for the alcoholysis reaction. The yields achieved were more than 90% in 2-MAGs production. Esterification of 2-MAGs produced with caprilic acid and lipases D and DF produced MLM structured lipids with 95% yield and 98% of caprilic acid at the extreme positions (Muñío et al. 2008, 2009, Esteban et al. 2009).

Two-steps process: Hydrolysis and Esterification

Hydrolysis reaction consists of converting a fat or ester decomposition into acids and alcohol in the presence of water. It is the inverse of esterification reaction (Fig. 9a).

Hydrolysis is the natural reaction catalyzed by lipases and it is used to obtain fatty acids from natural TAGs like vegetal oils (Rooney and Weatherley 2001), animal fats and, to a lesser extent, marine oils (Shimada et al. 1997b, 2001a).

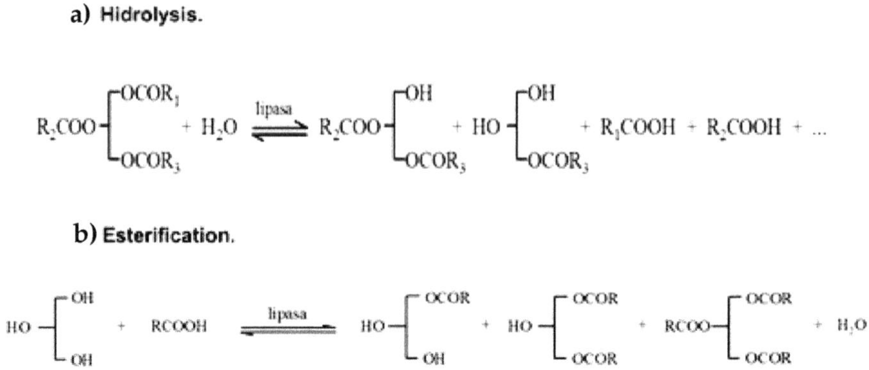

Fig. 9 Generic reactions of the two steps process (a) Hydrolysis and (b) Esterification.

As described earlier, esterification reaction happens between a fatty acid and an alcohol to produce an ester (Fig. 9b).

Because all reactions catalyzed by lipases consist in a hydrolysis reaction followed by an esterification reaction, it is possible to develope both reactions independently to optimize reaction conditions in each case (Willis and Marangoni 1999b). The mentioned authors carried out oil hydrolysis and partial glycerides esterification with Lipozyme IM®. After hydrolysis final products mixture was distillated to reduce free fatty acids content lower than 1%, so the composition was: 30% DAGs, 2% MAGs and 67% TAGs. In this work, the yield obtained by four ways were compared: acidolysis, interesterifiaction and combined hydrolysis-acidolysis-esterification and hydrolysis-interesterification.

The highest yield was obtained by the last method (83%), followed by interesterification (74%). Two ways of acidolysis produced lesser yields (60% in one step and 66% in the second step). This was because of a decrease in enzymatic activity due to pH decrease in the aqueous phase round to lipase or because of free fatty acids accumulation in oil-water interphase that prevented new substrate access to lipase. The main disadvantage of the two-step processes studied by the authors is the high DGAs content (around 17%) obtained with the final TAGs.

REFERENCES

Abbas, A.K., A.H. Lichtman and J.S. Pober. Cellular and Molecular Immunology, W.B. Saunders Company, Philadelphia, USA.

Ackman, R.G., W.M.N. Ratnayake and B. Olsson. 1988. The "basic" fatty acid composition of Atlantic fish oils: Potential similarities useful for enrichment of polyunsaturated fatty acids by urea complexation. J. Am. Oil Chem. Soc. 65: 136–138.

Ainsworth, S., C. Versteeg, M. Plamer and M.B. Millikan. 1996. Enzymatic interesterification of fats. Aust. J. Dairy Technol. 51: 105–107.

Akoh. C.C. 1995. Structured lipids-enzymatic approach. Inform. 6: 1055–1061.

Arvanitoyannis, I.S. and A. Kassaveti. 2008. Fish industry waste: Treatments, environmental impacts, current and potential uses. Int. J. Food Sci. Technol. 43: 726–745.

Augustsson, K., D.S. Michaud, E.B. Rimm, M.F. Leitzmann, M.J. Stampfer, W.C. Willett and E. Giovannucci. 2003. A prospective study of intake of fish and marine fatty acids and prostate cancer. Cancer Epidemiol. Biomarkers Prevent. 12: 64–67.

AWARENET. 2004. Handbook for the prevention and minimization of waste and valorization of by-products in European agro-food industries. Agro-food waste minimization and reduction network (AWARENET). Grow Programme, European Commision 1–7.

Babayan, V.K. 1987. Medium chain triglycerides and structured lipids. Lipids 22–6: 417–420.

Balk, E., M. Chung, A. Lichtenstein, P. Chew, B. Kupelnick, A. Lawrence, D. DeVine and J. Lau. 2004. Effects of omega-3 fatty acids on cardiovascular risk factors and intermediate markers of cardiovascular disease. Evidence report/technology assessment (Summary) 93: 1–6.

Bernadier, C.D. 1995. Advanced Nutrition: Macronutrients. CRC Press, Inc. Boca Ratón, Florida. USA.

Borum, P.R. 1992. Medium-chain triglycerides in formula for preterm neonates: implications for hepatic and extrahepatic metabolism. J. Pediatr. 120: 139–145.

Bourre, J.M., O. Dumont, M. Piciotti, M. Clément, J. Chaudière, M. Bonneil, G. Nalbone, H. Lafont, G. Pascal and G. Durand. 1991. Essentiality of omega 3 fatty acids for brain structure and function. World Rev. Nutr. Diet. 66: 103–117.

Breivik, H., G.G. Haraldsson and B. Kristinsson. 1997. Preparation of highly purified concentrates of eicosapentaenoic acid and docosahexaenoic acid. JAOCS, J. Am. Oil Chem. Soc. 74: 1425–1429.

Breivik, H. 2005. Concentrates. *In:* H. Breivik [eds.]. Long-Chain Omega-3 specialty oils. The oily Press, Bridgwater, England pp. 111–137.

Browning, L.M., J.D. Krebs, C.S. Moore, G.D. Mishra, M.A. O'Connell and S.A. Jebb. 2007. The impact of long chain n-3 polyunsaturated fatty acid supplementation on inflammation, insulin sensitivity and CVD risk in a group of overweight women with an inflammatory phenotype. Diabetes Obes. Metab. 9: 70–80.

Brunner, G. 1994. Gas extraction. An Introduction to Fundamentals of Supercritical Fluids and the Application to Separation Processes. Steinkopff Darmstadt Springer, New York, USA.

Camacho, F., A. Robles, B. Camacho, P.A. González, L. Esteban and E. Molina. 2007. Modeling of the kinetic for the acidolisis of different triacylglycerols and caprylic acid catalyzed by Lipozyme IM immobilized in a packed bed reactor. Chem. Eng. Sci. 62: 3127–3141.

Camacho Páez, B., A. Robles Medina, F. Camacho Rubio, P. González Moreno and E. Molina Grima. 2002. Production of structured triacylglycerides rich in n-3 polyunsaturated fatty acids by the acidolisis of cod liver oil and caprylic acid in a packed bed reactor: equilibrium and kinetics. Chem. Eng. Sci. 57–8: 1237–1249.

Carvalho, P., P. Campos, M. Noffs and J.G. Oliveira. 2003. Aplicacoes de lipases microbianas na obtencao de concentrados de acidos graxos poliinsaturados. Quim. Nova. 26: 75–80.

Carvalho, P., P. Campos, M. Noffs, P. Fregolente and L. Fregolente. 2009. Enzymatic hydrolysis of salmon oil by native lipases: Optimization of process parameters. J. Br. Chem. Soc. 20: 117–124.

Caughey, G.E., E. Mantzioris and R.A. Gibson. 1996. The effect on human tumor necrosis factor alpha and interleukin 1 beta production of diets enriched in n-3 fatty acids from vegetable oil or fish oil. Am. J. Clin. Nutr. 63: 116–122.

Christensen, M.S., C.E. Hoy, C.C. Becker and T.G. Redgrave. 1995. Intestinal absorption and lymphatic transport of eicosapentaeonic (EPA), docosahexaeonic (DHA); and decanoic acids: dependence on intramolecular triacylglycerol structure. Am. J. Clin. Nutr. 61: 56–61.

Crexi, V.T., M.L. Monte and A.A. Pinto. 2012. Polyunsaturated fatty acid concentrates of carp oil: Chemical hydrolysis and urea complexation. J. Am. Oil Chem. Soc. 89: 329–334.

Dullemeijer, C., J. Durga, I.A. Brouwer, O. Van De Rest, F.J. Kok, R.J.M. Brummer, M.P.J. Van Boxtel and P. Verhoef. 2007. N-3 Fatty acid proportions in plasma and cognitive performance in older adults. Am. J. Clin. Nutr. 86: 1479–1485.

Dunforda, N.T., M. Goto and F. Temelli. 1998. Modeling of oil extraction with supercritical CO_2 from Atlantic mackerel (*Scomber scombrus*) at different moisture contents. J. Supercrit. Fluids. 13: 303–309.

Dunstan, J.A. and S.L. Prescott. 2007. Does fish oil supplementation in pregnancy reduce the risk of allergic disease in infants? Curr. Opin. Allergy Clin. Immunol. 5: 215–221.

Eisenbach, W. 1984. Supercritical fluid extraction: a film demonstration. Berichte der Bunsengesellschaft/Phys. Chem. Chem. Phys. 88: 882–887.

Erkkilä, A.T., N.R. Matthan, D.M. Herrington and A.H. Lichtenstein. 2006. Higher plasma docosahexaenoic acid is associated with reduced progression of coronary atherosclerosis in women with CAD. J. Lipid Res. 47(12): 2814–2819.

Esquível, M.M., N.M. Bandarra, I. Fontan, M.G. Bernardo-Gil, I. Batista, M.L. Nunes and J.A. Empis. 1997. Supercritical carbon dioxide extraction of sardine Sardina pilchardus oil. LWT-Food Sci. Technol. 30: 715–720.

Esteban Cerdán, L., A. Robles Medina, A. Giménez Giménez, M.J. Ibáñez González and E. Molina Grima. 1998. Synthesis of polyunsaturated fatty acid-enriched triglycerides by lipase-catalized esterification. J. Am. Oil Chem. Soc. 75: 1329–1337.

Esteban, L., M.M. Muñío, A. Robles, E. Hita, M.J. Jiménez, P.A. González, B. Camacho and E. Molina. 2009. Synthesis of 2-monoacylglycerols (2-MAG) by enzymatic alcoholysis of fish oil using different reactor types. Biochem. Eng. J. 44: 271–279.

Falch, E., T. Rustad, R. Jonsdottir, N.B. Shaw, J. Dumay, J.P. Berge, S. Arason, J.P. Kerry, M. Sandbakk and M. Aursand. 2006. Geographical and seasonal differences in lipid composition and relative weight of by-products from gadiform species. J. Food Compos. Anal. 19: 727–736.

FAO. 2012. Global Production Statistics 1950–2010. Available at: http://www.fao.org.

Fauconnot, L., F. Robert, R. Villardand F. Dionisi. 2005. Chemical synthesis NMR characterization of structured polyunsaturated triacylglycerols. Chem. Phys. Lipids 139–2: 125–136.

Fedačko, J., D. Pella, V. Mechírová, P. Horvath, R. Rybár, P. Varjassyová and V. Vargová. 2007. N-3 PUFAs—From dietary supplements to medicines. Pathophysiology 14: 127–132.

Ferraro, V., I.B. Cruz, R.F. Jorge, F.X. Malcata, M.E. Pintado and P.M.L. Castro. 2010. Valorisation of natural extracts from marine source focused on marine by-products: A review. Food Res. Int. 43: 2221–2233.

Fitch Haumann, B. 1997. Structured lipids allow fat. Inform. 8: 1004–1011.

Fomuso, L.B. and C.C. Akoh. 1998. Structured lipids: lipase-catalized interesterificaction of tricaproin and trilinolein. J. Am. Oil Chem. Soc. 75: 405–410.

Frankel, E.N. 2005. Free Radical Oxidation. *In:* E.N. Frankel [eds.]. Lipid oxidation. The oily Press, Bridgwater, UK pp. 15–23.

Freeman, M.P., J.R. Hibbeln, K.L. Wisner, J.M. Davis, D. Mischoulon, M. Peet, P.E. Keck Jr., L.B. Marangell, A.J. Richardson, J. Lake and A.L. Stoll. 2006. Omega-3 fatty acids: Evidence basis for treatment and future research in psychiatry. J. Clin. Psychiat. 67 (12): 1954–1967.

Gámez-Meza, N., J.A. Noriega-Rodríguez, L.A. Medina-Juárez, J. Ortega-García, J. Monroy-Rivera, F.J. Toro-Vázquez, H.S. García and O. Angulo-Guerrero. 2003. Concentration of eicosapentaenoic acid and docosahexaenoic acid from fish oil by hydrolysis and urea complexation. Food Res. Int. 36: 721–727.

Gandhi, N.N. and K.D. Mukherjee. 2001. Synthesis of designed lipids using papaya (*Carica papaya*) latex lipase. J. Molecular Catalysis B: Enzymatic 11–4: 271–277.

Garg, M.L., A.B.R. Thomson and M. T. Clandinin. 1990. Interactions of saturated, n-6 and n-3 polyunsaturated fatty acids to modulate arachidonic acid metabolism. J. Lipid Res. 31: 271–277.

Geelen, A., J.M. Schouten, C. Kamphuis, B. E. Stam, J. Burema, J.M.S. Renkema, E.J. Bakker, P. Van't Veer and E. Kampman. 2007. Fish consumption, n-3 fatty acids, and colorectal cancer: A meta-analysis of prospective cohort studies. Am. J. Epidemiol. 166: 1116–1125.
Gildberg, A., J.A. Arnesen and M. Carlehög. 2002. Utilisation of cod backbone by biochemical fractionation. Process Biochem. 38: 475–480.
González Moreno, P.A., A. Robles Medina, F. Camacho Rubio, B. Camacho Páez and E. Molina Grima. 2004. Production of Structured Lipids by Acidolysis of an EPA-Enriched Fish Oil and Caprylic Acid in a Packed Bed Reactor: Analysis of Three Different Operation Modes. Biotechnol. Progress 20–4: 1044–1052.
Grigorakis, K., M.N. Alexis, K.D.A. Taylor and M. Hole. 2002. Comparison of wild and cultured gilthead sea bream (Sparus aurata); composition, appearance and seasonal variations. Int. J. Food Sci. Technol. 37: 477–484.
Halldorsson, A., C.D. Magnusson and G.G. Haraldsson. 2003. Chemoenzymatic synthesis of structured triacylgycerols. Tetrahedron Letters 42–43: 7675–7677.
Han, J.J., Y. Iwasaki and T. Yamane. 1999. Monitoring of lipase-catalyzed transesterification between eicosapentaenoic acid ethyl ester and tricaprylin by silver ion High-Performance Liquid Chromatography and High-Temperature Gas Chromatography. J. Am. Oil Chem. Soc. 76: 31–39.
Haraldsson, G. 1983. Separation of saturated/unsaturated fatty acids. J. Am. Oil Chem. Soc. 61: 219–222.
Haraldsson, G.C., P. A. Höskuldsson, S.Th. Sigurdsson, F. Thorsteinsson and S. Gudbjarnason. 1989. The preparation of triglycerides highly enriched with w-3 polyunsaturated fatty acids via lipase catalyzed interesterification. Tethahedron Lett. 30: 1671–1674.
Haraldsson, G.G., B. Kristinsson, R. Sigurdardottir, G.G. Gudmundsson and H. Breivik. 1997. The preparation of concentrates of eicosapentaenoic acid and docosahexaenoic acid by lipase-catalyzed transesterification of fish oil with ethanol. J. Am. Oil Chem. 74: 1419–1424.
Harris, W.S., M. Miller, A.P. Tighe, M.H. Davidson and E.J. Schaefer. 2008. Omega-3 fatty acids and coronary heart disease risk: Clinical and mechanistic perspectives. Atherosclerosis. 197: 12–24.
Hayes, D.G., Y.C. Bengtsson, J.M. Van Alstine and F. Setterwall. 1998. Urea complexation for the rapid, ecologically responsible fractionation of fatty acids from seed oil. J. Am. Oil Chem. Soc. 75: 1403–1409.
Healy, D.A., F.A. Wallace and E.A. Miles. 2000. Effect of low-to-moderate amounts of dietary fish oil on neutrophil lipid composition and function. Lipids 35: 763–768.
Helland, I.B., L. Smith, K. Saarem, O.D. Saugstad and C.A. Drevon. 2003. Maternal supplementation with very-long.chain n-3 fatty acids during pregnancy and lactation augments children's IQ at 4 years of age. Pedriatric 111: 39–44.
Hita, E., A. Robles, B. Camacho, A. Ramírez, L. Esteban, M.J. Jiménez, M.N. Muñío, P.A. González and E. Molina. 2007. Production of structured triacylglycerols (STAG) rich in docosahexaenoic acid (DHA) in position 2 by acidolisis of tuna oil catalyzed by lipases. Process Biochem. 42: 415–422.
Hita, E., A. Robles, B. Camacho, P.A. González, L. Esteban, M.J. Jiménez, M.M. Muñío and E. Molina. 2009. Production of structured triacylglycerols by acidolysis catalyzed by lipases immobilized in a packed bed reactor. Biochem. Eng. J. 46: 257–264.
Hu, F.B., E. Cho, K.M. Rexrode, C.M. Albert and J. E. Manson. 2003. Fish and long-chain ω-3 fatty acid intake and risk of coronary heart disease and total mortality in diabetic women. Circulation 107(14): 1852–1857.
Innis, S.M., R. Dyer, P. Quinlan and D. Diersen-Schade. 1995. Palmitic acid is absorbed as sn-2 monopalmitin from milk and formula with rearranged triacylglycerols and results in increased plasma triglyceride sn-2 and cholesterol ester palmitate in piglets. J. Nutr. Vol. 125: 73–81.

Irimescu, R., M. Yasui, Y. Iwasaki, N. Shimidzu and T. Yamane. 2000. Enzymatic synthesis of 1, 3-dicapryoyl-2-eicosapentaenoylglycerol. J. Am. Oil Chem. Soc. 77: 501–506.

Irimescu, R., K. Furihata, K. Hata, Y. Iwasaki and T. Yamane. 2001a. Utilization of reaction medium-dependent regiospecificity of *Candida antarctica* (Novozym 435) for the synthesis of 1, 3-dicapryloyl-2-docosahexaenoyl (or eicosapentaenoyl) glycerol. J. Am. Oil Chem. Soc. 78–3: 285–289.

Irimescu, R., K. Furihata, K. Hata, Y. Iwasaki and T. Yamane. 2001b. Two-step enzymatic synthesis of docosahexaenoic acid-rich symmetrically structured triacylglycerols via 2-monoacylglycerols. J. Am. Oil Chem. Soc. 78–7: 743–748.

Jennings, B.H. and C.C. Akoh. 2001. Lipase catalyzed modification of fish oil to incorporate capric acid. Food Chemistry 72–3: 273–278.

Kim, S.K. and E. Mendis. 2006. Bioactive compounds from marine processing byproducts—A review. Food Res. Int. 39: 383–393.

Kojima, Y., E. Sakuradani and S. Shimizu. 2006. Different specificities of two types of *Pseudomonas* lipases for C20 fatty acids with a $\Delta5$ unsaturated double and their application for selective concentration of fatty acids. Different specificities of two types of *Pseudomonas* lipases for C20 fatty acids with a $\Delta5$ unsaturated double and their application for selective concentration of fatty acids. J. Biosci. Bioeng. 1: 490–500.

Kristinsson, H.G. and B.A. Rasco. 2000. Fish protein hydrolysates: Production, biochemical, and functional properties. Critical Reviews in Food Science and Nutrition 40(1): 43–81.

Laerum, B.N., C. Svanes, T. Wentzel-Larsen, A. Gulsvik, K. Torén, E. Norrman, T. Gíslason, C. Janson and E. Omenaas. 2007. Young maternal age at delivery is associated with asthma in adult offspring. Respiratory Medicine 101:1431–1438.

Létisse, M. and L. Comeau. 2008. Enrichment of eicosapentaenoic acid and docosahexaenoic acid from sardine by-products by supercritical fluid fractionation. J. Sep. Sci. 31: 1374–1380.

Liaset, B., E. Lied and M. Espe. 2000. Enzymatic hydrolysis of by-products from the fish-filleting industry; chemical characterisation and nutritional evaluation. J. Sci. Food Agric. 80: 581–589.

Linko, R.R. and H. Karinkanta. 1970. Fatty acids of long chain length in baltic herring lipids. J. Am. Oil Chem. Soc. 47: 42–46.

Liu, S., C. Zhang, P. Hong and H. Ji. 2006. Concentration of docosahexaenoic acid (DHA) and eicosapentaenoic acid (EPA) of tuna oil by urea complexation: Optimization of process parameters. J. Food Eng. 73: 203–209.

López-Martínez, J.C., P. Campra-Madrid and J.L. Guil-Guerrero. 2004. γ-linolenic acid enrichment from Borago officinalis and Echium fastuosum seed oils and fatty acids by low temperature crystallization. J. Biosci. Bioeng. 97: 294–298.

Lyberg, A.M. and P. Adlercreutz. 2008. Lipase specificity towards eicosapentaenoic acid and docosahexaenoic acid depends on substrate structure. Biochim. Biophys. Acta. 1784: 343–350.

Magrum, L.J. and P.V. Johnston. 1983. Modulation of prostaglandin synthesis in rat peritoneal macrophages with n-3 fatty acids. Lipids 18: 514–521.

Malik, M., A.J. Camm, J.T. Bigger Jr., G. Breithardt, S. Cerutti, R.J. Cohen, P. Coumel, E.L. Fallen, H.L. Kennedy, R.E. Kleiger, F. Lombardi, A. Malliani, A.J. Moss, J.N. Rottman, G. Schmidt, P.J. Schwartz and D.H. Singer. 1996. Heart rate variability. Standards of measurement, physiological interpretation, and clinical use. Eur. Heart J. 17: 354–381.

Marchioli, R., R.M. Marfisi, G. Borrelli, C. Chieffo, M.G. Franzosi, G. Levantesi, A.P. Maggioni, G.L. Nicolosi, M. Scarano, M.G. Silletta, C. Schweiger, L. Tavazzi and G. Tognoni. 2007. Efficacy of n-3 polyunsaturated fatty acids according to clinical characteristics of patients with recent myocardial infarction: Insights from the GISSI-Prevenzione trial. J. Cardiovasc. Med. 8 (Suppl. 1): S34–S37.

Martín, A. and M.J. Cocero. 2007. Mathematical modeling of the fractionation of liquids with supercritical CO_2 in a countercurrent packed column. J. Superc. Fluids. 39: 304–314.

Martins, P.F., C. B. Baristella, R. Maciel Filho and M.R. Wolf-Macie. 2006. Comparison of two different strategies for tocopherols enrichment using a molecular distillation process. Ind. & Eng. Chem. Res. 45: 753.
Mbatia, B., P. Adlercreutz, F. Mulaa and B. Mattiasson. 2010. Enzymatic enrichment of n-3 polyunsaturated fatty acids in Nile perch (Lates niloticus) viscera oil. Eur. J. Lipid Sci. Technol. 112: 977–984.
Mbatia, B., B. Mattiasson, F. Mulaa and P. Adlercreutz. 2011. Strategies for the enzymatic enrichment of PUFA from fish oil. Eur. J. Lipid Sci. Technol. 113: 717–723.
Meher, L.C., D.V. Sagar and S.N. Naik. 2006. Technical aspects of biodiesel production by transesterification—a review. Renew. Sustain. Energy Rev. 10: 248–268.
Merolli, A. 1997. Medium-chain lipids: new sources, uses. Inform. 8: 597–603.
Miyake, Y., S. Sasaki, M. Arakawa, K. Tanaka, K. Murakami and Y. Ohya. 2008. Fatty acid intake and asthma symptoms in Japanese children: The Ryukyus child health study. Clin. Exp. Allergy 38: 1644–1650.
Moore, C.S., S.P. Bryant, G.D. Mishra, J.D. Krebs, L.M. Browning, G.J. Miller and S.A. Jebb. 2006. Oily fish reduces plasma triacylglycerols: a primary prevention study in overweight men and women. Nutr. 22: 1012–1024.
Mori, T.A., V. Burke, I.B. Puddey, G.F. Watts, D.N. O'Neal, J.D. Best and L.J. Beilin. 2000. Purified eicosapentaenoic and docosahexaenoic acids have differential effects on serum lipids and lipoproteins, LDL particle size, glucose, and insulin in mildly hypedipidemic men. Am. J. Clin. Nutr. 71: 1085–1094.
Morris, M.C., D.A. Evans, C.C. Tangney, J.L. Bienias and R.S. Wilson. 2005. Fish consumption and cognitive decline with age in a large community study. Arch. Neurol. 62: 1849–1853.
Muñío, M.M., L. Esteban, A. Robles, E. Hita, M.J. Jiménez, P.A. González, B. Camacho and E. Molina. 2008. Synthesis of 2-monoacylglycerols rich in polyunsaturated fatty acids by ethanolysis of fish oil catalyzed by 1, 3 specific lipases. Process Biochem. 43: 1033–1039.
Muñío, M.M., A. Robles, L. Esteban, P.A. González and E. Molina. 2009. Synthesis of structured lipids by two enzymatic steps: ethanolysis of fish oils and esterification of 2-monoacylglycerols. Process Biochem. 44: 723–730.
Nair, S.S.D.J.W. Leitch, J. Falconer and M.L. Garg. 1997. Prevention of cardiac arrhythmia by dietary (n-3) polyunsaturated fatty acids and their mechanism of action. J. Nutr. 127: 383–393.
Nilsson, W.B., E.J. Gauglitz Jr. and J. K. Hudson. 1989. Supercritical fluid fractionation of fish oil esters using incremental pressure programming and a temperature gradient. J. Am. Oil Chem. Soc. 66: 1596–1600.
Nkondjock, A. and O. Receveur. 2003a. Fish-seafood consumption, obesity, and risk of type 2 diabetes: An ecological study. Diabetes Metab. 29: 635–642.
Nkondjock, A. and B. Shatenstein, P. Maisonneuve and P. Ghadirian. 2003b. Specific fatty acids and human colorectal cancer: An overview. Cancer Detect. Prev. 27: 55–66.
Okada, T. and M.T. Morrissey. 2008. Production of n-3 polyunsaturated fatty acid concentrate from sardine oil by immobilized Candida rugosa lipase. J. Am. Oil Chem. 73: C146–C150.
Oliveira, A.C.M. and P.J. Bechtel. 2005. Lipid composition of Alaska Pink Salmon (*Oncorhynchus gorbuscha*) and Alaska Walleye Pollock (*Theragra chalcogramma*) by-products. J. Aquat. Food Prod. Technol. 14: 73–91.
Parker, G., N.A. Gibson, H. Brotchie, G. Heruc, A.M. Rees and D. Hadzi-Pavlovic. 2006. Omega-3 fatty acids and mood disorders. Am. J. Psychiat. 163: 969–978.
Pearson, A.M., J.D. Love and F.B. Shorlan. 1977. "Warmed-Over" Flavor in Meat, Poultry and Fish. Adv. Food Res. 23: 1–74.
Perretti, G., A. Motori, E. Bravi, F. Favati, L. Montanari and P. Fantozzi. 2007. Supercritical carbon dioxide fractionation of fish oil fatty acid ethyl esters. J. Supercr. Fluids 40: 349–353.

Rai, A.K., H.C. Swapna, N. Bhaskar and V. Baskaran. 2011. Potential of seafood industry by-products as sources of recoverable lipids: fatty acid composition of meat and nonmeat component of selected Indian marine fishes. J. Food Biochem. 36. 441–448.

Rallidis, L.S., G. Paschos, G.K. Liakos, A. H. Velissaridou, G. Anastasiadis and A. Zampelas. 2003. Dietary α-linolenic acid decreases C-reactive protein, serum amyloid A and interleukin-6 in dyslipidaemic patients. Atherosclerosis 167(2): 237–242.

Rooney, D. and L.R. Weatherley. 2001. The effect of reaction conditions upon lipase catalysed hydrolysis of high oleate sunflower oil in a stirred liquid-liquid reactor. Process Biochem. 36-10: 947–953.

Rose, D.P. and J.M. Connolly. 1999. Omega-3 fatty acids as cancer chemopreventive agents. Pharmacol. Therapeut. 83: 217–244.

Rubio, N. and S. Beltrán. 2011. Supercritical fluid technology for extraction, concentration & formulation of Omega-3 rich oils. A novel strategy for valorization of fish by-products. PhD Thesis, University of Burgos.

Ruxton, C.H.S., P.C. Calder, S.C. Reed and M.J.A. Simpson. 2005. The impact of long-chain n-3 polyunsaturated fatty acids on human health. Nutr. Res. Rev. 18: 113–129.

Sahena, F., I.S.M. Zaidul, S. Jinap, N. Saari, H. A. Jahurul, K.A. Abbas and N.A. Norulaini. 2009. PUFAs in fish: Extraction, fractionation, importance in health. Compr. Rev. Food Sci. Food Saf. 8: 59–74.

Saito, H., Y. Seike, H. Ioka, K. Osako, M. Tanaka, A. Takashima, J.M. Keriko, S. Kose and J.C. Rodriguez Souza. 2005. High docosahexaenoic acid levels in both neutral and polar lipids of a highly migratory fish: *Thunnus tonggol* (Bleeker). Lipids 40: 941–953.

Salah, S. and S. Saloua. 2010. The influence of season on lipid content and fatty acids profile of *Euthynnus alletteratus* fillets and by-products. J. Muscle Foods 21: 365–378.

Sathivel, S. 2011. Fish oil extraction, purification, and its properties. *In*: C. Alasalvar, F. Shahidi, K. Miyashita and U. Wanasundara [eds.]. Handbook of Seafood Quality, Safety and Health Applications.Wiley-Blackwell, West-Sussex. United Kingdom pp. 423–430.

Schaefer, E.J., V. Bongard, A.S. Beiser, S. Lamon-Fava, S.J. Robins, R. Au, K.L. Tucker, D.J. Kyle, P.W.F. Wilson and P.A. Wolf. 2006. Plasma phosphatidylcholine docosahexaenoic acid content and risk of dementia and alzheimer disease: The framingham heart study. Arch. Neurol. 63: 1545–1550.

Schnappinger, M., S. Sausenthaler, J. Linseisen, H. Hauner and J. Heinrich. 2009. Fish consumption, allergic sensitisation and allergic diseases in adults. Ann. Nutr. Metab. 54: 67–74.

Serhan, C.N. and J. Savill. 2005. Resolution of inflammation: The beginning programs the end. Nature Immun. 6(12): 1191–1197.

Shahidi, F. and Y. Zhong. 2010. Lipid oxidation and improving the oxidative stability. Chem. Soc. Rev. 39: 4067–4079.

Shimada, Y., A. Sugihara, H. Nakano, T. Yolota, T. Nagao, S. Komemushi and Y. Tominaga. 1996a. Production of structured lipids containing essential fatty acids by immobilized *Rhizopus delemar* lipase. J. Am. Oil Chem. Soc. 73: 1415–1420.

Shimada, Y., A. Sugihara, K. Maruyama, T. Nagao, S. Nakayama, H. Nakano and Y. Tominaga. 1996b. Production of structured lipids containing docosahexaenoic and caprylic acids using immobilized *Rhizopus delemar* lipase. J. Ferment. Bioeng. 81: 299–303.

Shimada, Y., A. Sugihara, H. Nakano, T. Kuramoto, T. Nagao, M. Gemba and Y. Tominaga. 1997a. Purification of docosahexaenoic acid by selective esterification of fatty acids from tuna oil with *Rhizopus delemar* lipase. J. Am. Oil Chem. Soc. 74: 97–101.

Shimada, Y., K. Maruyama, A. Sugihara, S. Moriyama and Y. Tominaga. 1997b. Purification of docosahexaenoic acid from tuna oil by two step enzymatic method: hydrolisys and selective esterification. J. Am. Oil Chem. Soc. 74: 1441–1446.

Shimada, Y., A. Sugihara and Y. Tominaga. 2001a. Enzymatic purification of polyunsaturated fatty acids. J. Biosci. Bioeng. 91: 529–538.

Shimada, Y., Y. Watanabe, A. Sugiara, T. Baba, T. Ooguri, S. Moriyama, T. Terai and Y. Tominaga. 2001b. Ethyl esterification of docosahexaenoic acid in an organic solvent-free system with immobilized *Candida antarctica* lipase. J. Biosci. Bioeng. 92: 19–23.
Skaliotis, L. 2011. Short path evaporation for the production of premium quality oils. Lipid Technol. 23: 204–206.
Small, D.M. 1991. The effects of glyceride structure on absorption and metabolism. Annu. Rev. Nutr. 11: 413–434.
Soumanou, M.M., U.T. Bornscheuer and R.D. Schmid. 1998a. Two-step enzymatic reaction for the synthesis of pure structured triacylglycerides. J. Am. Oil Chem. Soc. 75–6: 703–710.
Soumanou, M.M., U.T. Bornscheuer, U. Schmid and R.D. Schmid. 1998b. Synthesis of structured triglycerides by lipase catalysis. Fett/Lipid. 100-4-5: 156–160.
Tanaka, Y., J. Hirano and T. Funada. 1992. Concentration of docosahexaenoic acid in glyceride by hydrolysis of fish oil with Candida cylindricae lipase. J. Am. Oil Chem. Soc. 69: 1210–1214.
Tanaka, Y., T. Funada, J. Hirano and R. Hashizume. 1993. Triglyceride specificity of Candida cylindracea lipase-effect of docosahexaenoic acid on resistance of triglycerides to lipase. J. Am. Oil Chem. 70: 1031–1034.
Tarpila, S., A. Aro and I. Salminen. 2002. The effect of flaxseed supplementation in processed foods on serum fatty acids and enterolactone. Eur. J. Clin. Nutr. 56: 157–165.
Terry, P., P. Lichtenstein, M. Feychting, A. Ahlbom and A.Wolk. 2001. Fatty fish consumption and risk of prostate cancer. Lancet. 357 (9270): 1764–1766.
Thorsdottir, I., H. Tomasson, I. Gunnarsdottir, E. Gisladottir, M. Kiely, M.D. Parra, N.M. Bandarra, G. Schaafsma and J.A. Martinéz. 2007. Randomized trial of weight-loss-diets for young adults varying in fish and fish oil content. Int. J. Obesity. 31(10): 1560–1566.
Tidwell, J.H. and G.L. Allan. 2001. Fish as food: Aquaculture's contribution. Ecological and economic impacts and contributions of fish farming and capture fisheries. EMBO Reports 2: 958–963.
Torres, C.F., T.J. Nettekoven and C.G. Hill, Jr. 2003. Preparation of purified acylglycerols of eicosapentaenoic acid and docosahexaenoic acid and their re-esterification with conjugated linolenic acid. Enzyme Microb. Technol. 32: 49–58.
Torres, J.A., Y.C. Chen, J. Rodrigo-García and J. Jaczyznski. 2007. Recovery of by-products from seafood processins streams. *In*: F. Shahidi [ed.]. Maximising the value of marine by-products. Woodhead Publishing Ltd., Cambridge, UK pp. 65–90.
Wanasundara, U.N. and F. Shahidi. 1998. Lipase-assisted concentration of n-3 polyunsaturated fatty acids in acylglycerols from marine oils. J. Am. Oil Chem. Soc. 75: 945–951.
Wijesundera, R.C., W.M.N. Ratnayake and R.G. Ackman. 1989. Eicosapentaenoic acid geometrical isomer artifacts in heated fish oil esters. J. Am. Oil Chem. Soc. 66: 1822–1830.
Willis, W.M. and A.G. Marangoni. 1999a. Assessment of lipase and chemically catalyzed lipid modification strategies for the production of structured lipids. J. Am. Oil Chem. Soc. 76: 443–450.
Willis, W.M. and A.G. Marangoni. 1999b. Biotechnological strategies for the modification of food lipids. Biotechnol. Gen. Eng. Rev. 16: 141–175.
Xu, X., S. Balchen, C.E. Høy and J. Adler-Niessen. 1998. Pilot batch production of specific-structured lipids by lipase catalyzed interesterification: preliminary study on incorporation and acyl migration. J. Am. Oil Chem. Soc. 75: 301–308.
Xu, X., L.B. Fomuso and C.C. Akoh. 2000a. Synthesis of structured triacylglycerols by lipase catalyzed acidolisis in a packed bed bioreactor. J. Agric. Food Chem. 48-1: 3–10.
Xu, X., L.B. Fomuso and C.C. Akoh. 2000b. Modification of menhaden oil by enzymatic acidolisis to produce structured lipids: optimization by response surface design in a packed bed reactor. J. Am. Oil Chem. Soc. 77-2: 171–176.

Yamane, T., S. Tomomasa, Y. Sahashi, L. Vikersveen and T. Hocino. 1992. Production of n-3 PUFA-enriched fish oil by lipase-catalyzed acidolysis without solvent. J. Am. Oil Chem. Soc. 69: 1104–1107.

Yamane, T., T. Suzuki and T. Hocino. 1993. Increasing n-3 polyunsaturated fatty acid content of fish oil by temperature control of lipase-catalyzed acidolysis. J. Am. Oil Chem. Soc. 70: 1285–1287.

Zaidul, I.S.M., N.A.N. Norulaini, A.K.M. Omar, Y. Sato and R.L. Smith Jr. 2007. Separation of palm kernel oil from palm kernel with supercritical carbon dioxide using pressure swing technique. J. Food Eng. 81: 419–428.

8

Biodiesel Production from Fish Oil

Mohriam Khanum,[1] *Pedro J. García-Moreno,*[2,*] *Antonio Guadix*[2] and *Emilia M. Guadix*[2]

ABSTRACT

The high cost of biodiesel produced from vegetable oils and the problems of land space, due to the food vs. fuel issue, stimulate the need to find a cheaper alternative feedstock for biodiesel production. Fish oil may have the possibility to solve these problems, particularly when it is produced from wastes generated by the fish processing industry.

In this chapter, it is evaluated the feasibility to produce biodiesel from salmon oil employing a two-step transesterification process with methanol and sodium hydroxide as catalyst. The quality of the biodiesel produced was investigated, in terms of acid value, FAME content, biodiesel yield, glycerine mass, viscosity, density, oxidative stability and thermal properties, at different operating conditions, temperature (40°C, 50°C and 60°C), methanol to oil ratio (3:1, 6:1 and 9:1) and reaction time (30 min, 60 min and 90 min).

[1] Department of Chemical Engineering, University College London, Gower Street, WC1E 6BT London, United Kingdom.
Email: mkhanum@ucl.ac.uk

[2] Department of Chemical Engineering, University of Granada, Severo Ochoa, s/n, 18071 Granada, España.

* Corresponding author: pjgarcia@ugr.es

INTRODUCTION

The most common feedstock used in the production of biodiesel are vegetable oils, however, cultivated land is today limited to grow enough seed-oil plants. Due to the indirect land use and food vs. fuel issue, the utilization of these feedstocks for biodiesel production has been criticized (Austic et al. 2010). Feedstock accounts for 70% of the price of biodiesel (University of Alaska Fairbanks 2006). Hence, there is a need to find cheaper feedstocks.

The fish processing industry generates large quantities of tissue waste and by-products, which is approximately 50% of the total processed fish. Although this waste is generally discarded or sold at a very low price as fertilizer or animal feed, fish oil can be obtained from the fish by-products which could be used as feedstock for biodiesel production. Converting fish oil into biodiesel avoids many engine problems as it has a reduced viscosity. In addition, biodiesel burns more cleanly than fish oil which contains impurities such as water and soluble proteins, resulting in more particulate laden emissions and carbon deposits on the fuel injectors and in the engine (University of Alaska Fairbanks 2006).

On the other hand, fish oil contains a high percentage of omega-3 PUFAs. Because of this, it has a higher commercial demand and value, due to the human health benefits of omega-3. Consequently, it would not be suitable to use fish oil high in PUFAs for the production of biodiesel. As salmon oil contains a lower amount of PUFAs as compared to other oily fish such as tuna and cod, it has a lower commercial demand and value and hence, it is more suitable for biodiesel production. Salmon processing industries also generates large quantities of by-products, which will be required if biodiesel production from salmon oil is to be commercialized and produced in large quantities. For these reasons salmon oil appears to be a more viable type of fish oil to produce biodiesel from and therefore, it has been employed for this research project.

Using waste fish oil as an alternative feedstock for biodiesel has the potential to reduce the production costs drastically. However, there is limited research on the fuel characteristics of biodiesel produced from crude fish oil. Previous research carried out on the production of biodiesel from fish oil has proved that fish oil is a viable option that can be used to produce biodiesel presenting good fuel properties. Previous research has also provided details about the difference and difficulties of producing biodiesel from fish oil when compared to the usual vegetable oils (Avena-Bustillos et al. 2007, Avena-Bustillos et al. 2008, Austic et al. 2010).

In this chapter, the fuel characteristics of biodiesel produced from salmon oil have been investigated further as well as the feasibility of using fish oil as feedstock to produce biodiesel presenting an appropriate quality.

Biodiesel

Biodiesel is a renewable alternative diesel fuel. It is environmentally friendly due to its biodegradable and nontoxic properties, as well as having low emissions profiles. Although biodiesel is most commonly used directly as fuel in automotive vehicle engines, it is miscible with petrodiesel in all ratios and therefore, it is also used in blends of biodiesel with petrodiesel. Biodiesel can also be used as heating oil, aviation fuel in low flying aircrafts and cleaning agents (Knothe et al. 2005a).

The history of biodiesel

The use of vegetable oil in diesel engines has been present for more than 100 years, the inventor of the diesel engine, Rudolph Diesel, used peanut oil in his prime model in Augsburg, Germany 1893. Although, the first use of vegetable oils as alternative fuels occurred at the Paris Exposition of 1900 where the Otto Company showed a diesel engine that ran completely on Arachide, peanut oil (Knothe et al. 2005a). In support of this Diesel conducted many successful experiments on vegetable oils as a fuel for diesel engines and during a speech in 1912 said "the use of vegetable oils for engine fuels may seem insignificant today but such oils may become, in the course of time, as important as petroleum and the coal-tar products of the present time". However, due to the cheap price of petroleum and the problems of using vegetable oil directly in diesel engines it was not commercialized (Knothe et al. 2005a).

During World War II vegetable oils were tested and used as diesel fuels by many countries including France, Germany, United Kingdom and China, despite the widespread use of fossil fuels. There were some operational problems mainly due to the high viscosity of the vegetable oils compared to petrodiesel. Heating vegetable oils, blending them with petrodiesel or ethanol pyrolysis and cracking of the oils were all methods used in an attempt to overcome these problems. In 1937 G. Chavanne of the University of Brussels was granted a patent for a "Procedure for the transformation of vegetable oils for their uses as fuels". The patent is the first account of the production of biodiesel. It describes the acid-catalysed transesterification of palm oil using ethanol, although methanol is mentioned, to be able to separate the fatty acids from glycerol (Knothe et al. 2005a).

By 1983 the process of producing fuel-quality, engine-tested biodiesel was completed and published internationally and in 1989 the first industrial scale biodiesel plant was constructed by an Austrian company, Gaskoks, with the capacity of 30,000 tonnes of rapeseed per annum. Further to this, France began local production of biodiesel blended with regular petrodiesel. By 1998, the Austrian Biofuels Institute had identified 21 countries with commercial biodiesel projects and 100% biodiesel was available at many service stations across Europe. Finally, in September 2005 Minnesota became the first U.S state to make it mandatory that all diesel fuel sold in that state must contain at least 2% biodiesel (Knothe et al. 2005a).

Currently, fossil fuels, gases and coal provide the majority of energy in the world. However, these sources of energy are rapidly depleting due to the rate of usage and they also have a significant negative impact on the environment. Hence, there is a need for alternative renewable fuels (Meher et al. 2006). Biodiesel has several advantages over petrodiesel in the technical aspect; however, it does have disadvantages compared to petrodiesel mainly concerning the economics of production. The advantages and disadvantages have been listed in Table 1 below (Knothe et al. 2005a).

From Table 1, it is clear that there are many advantages of biodiesel, which make it a key alternative to fossil fuels and providing solutions to the current problems associated with petrochemicals. However, the major disadvantage that biodiesel needs to overcome to be able to compete with petrol diesel is its high cost. The high cost of biodiesel is mainly associated with the feedstock; by using waste oils as an alternative feedstock this can be reduced.

Table 1 Advantages and Disadvantages of Biodiesel.

Advantages	Disadvantages
• Derived from a renewable domestic resource, hence reduced dependence on and preserves petroleum. • Reduction of greenhouse gases and lower harmful emissions. • Biodegradable. • Higher flash point hence safer handling and storage. • May improve rural economics due to agricultural surplus used as raw material. • Legislative and regulatory incentives/ subsidiaries such as reduced excise taxes in many countries. • Non-toxic. • Increased lubricity.	• Higher cost/price. • Increased nitrogen oxide emissions. • More rapid degradation due to oxidation and hydrolysis, which give rise to transportation and storage problems. • Less suitable to use at lower temperatures due to higher 'cloud and pour points'. • May create competition between food crops and oil crops hence the possibility for famine in poor countries and problems with global food supply.

Biodiesel production process

Biodiesel is composed of monoalkyl esters of long fatty acid chains, which can be produced from a variety of renewable biological resources, such as vegetable oils and animal fats as well as waste oils (Van Gerpen et al. 2005). Vegetable oils and animal fats contain triglycerides that when reacted with an alcohol, commonly methanol, in the presence of a catalyst, usually potassium hydroxide and sodium hydroxide, produce alkyl esters. This is the most common method used to produce biodiesel, which is known as transesterification (Van Gerpen et al. 2005). Other methods of biodiesel production include pyrolysis, dilution with hydrocarbons and emulsification (Demirbas and Karslioglu 2007). The biodiesel yield obtained in the transesterification reaction is influenced by various parameters such as alcohol type, alcohol to oil molar ratio, reaction temperature, reaction time, catalyst type, water content of the reactants and quantity of the reactants (Avena-Bustillos et al. 2007). The reaction pattern can be seen in Fig. 1 (Van Gerpen et al. 2005).

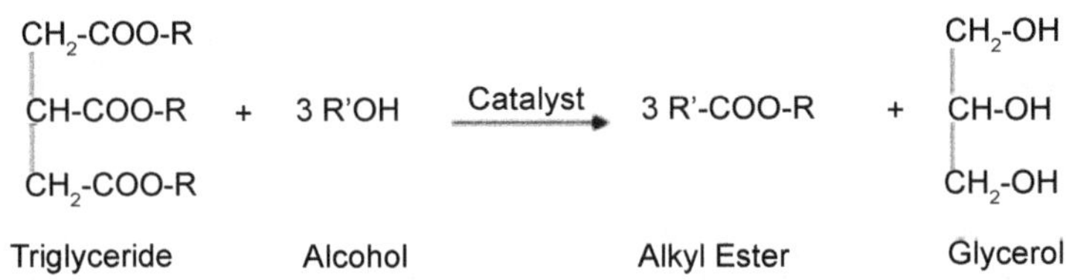

Fig. 1 Transesterification of Triglycerides.

Nevertheless, when the feedstock used to produce biodiesel presents a high free fatty acid content a two-step production process is required, as it is the most effective method to produce biodiesel with a high yield and rich in fatty acid methyl esters (Avena-Bustillos et al. 2007). The process involves two reaction stages: (i) the first uses an acid catalyst to convert the free fatty acids into esters before the alkali catalyst is used, avoiding the formation of soap that could decrease the biodiesel yield and complicate the separation and purification processes; (ii) the second uses the usual alkaline catalyst which is much faster than the acid-catalysed reaction (Jayaraj et al. 2004). After the two reaction stages are completed, the mono alkyl esters produced are purified and separated from the by-product glycerol and other impurities such as alcohol, mono- and di-glycerides and water. The process employed in this study is shown in the flow diagram in Fig. 2.

Biodiesel specifications

Biodiesel produced must meet the European Standard, EN 14124, to ensure the fuel quality. Some of the key fuel properties that contribute to the fuel quality are acid value, fatty acid methyl ester (FAME) content, viscosity, density and oxidative stability (Table 2). If the limits set by the standards for these properties are not met, they will have severe consequences on the engine and may cause operational problems.

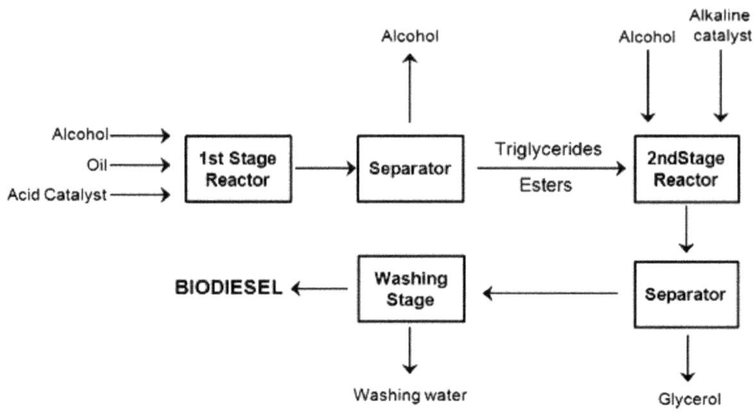

Fig. 2 Two-step Production Process for Biodiesel.

Table 2 Key Biodiesel Specifications, EN 14124.

Property	Minimum	Maximum
Acid value	-	0.50 mg KOH/g
FAME content	96.5%	-
Viscosity at 40°C	3.5 mm^2/s	5 mm^2/s
Density at 15°C	860 kg/m^3	900 kg/m^3
Oxidative stability at 110°C	6 hr	-

Other aspects to consider when producing diesel fuels are the performance problems caused when vehicles and fuel systems are subjected to cold temperatures. The cold flow properties most commonly used for biodiesel are cloud point and pour point. Low temperatures cause solid wax crystal nuclei to form, if the temperature decreases further these crystals grow in size. The temperature at which crystals become visible is defined as the cloud point as the crystals usually form a cloudy or hazy suspension (Knothe et al. 2005c). If the engine is operated at a temperature below the cloud point, it can result in fuel filter clogging due to the wax crystals. At temperatures below the cloud point, larger crystals fuse together and form large agglomerates that can restrict or cut off flow through fuel lines and filters. The temperature at which crystal agglomeration is extensive enough to prevent free pouring of fluid is defined as the "pour point" (Knothe et al. 2005c). The cold-filter plugging point and low-temperature flow test are methods employed to determine the operability limits of the fuel, as operability temperature limits cannot be accurately predicted using the cloud point and pour point measurements (Dunn 1999).

Biodiesel made from feedstocks containing higher concentrations of high-melting point saturated long-chain fatty acids tend to have relatively poor cold flow properties (Knothe et al. 2005c). Hence, it is one of the key disadvantages of biodiesel and must be resolved before biodiesel can become fully competitive with petrodiesel. Although, biodiesel produced from fish oils, due to its high polyunsaturated fatty acid content, present better cold flow properties than biodiesel produced from other feedstocks.

Fish Oil

Raw material used for the production of fish oil and fishmeal falls in three categories (The AOCs Lipid Library 2011): (i) fish caught specifically for the reduction to fish oil and fishmeal; (ii) fish caught incidentally or is a by-catch from another fishery, and (iii) fish by-products from edible fisheries such as cuttings from filleting operations, fish cannery waste, roe fishery waste and surimi processing waste. Generally fish oil is derived from the white fish offal such as cod and halibut, as well as small fish such as anchovy and capelin or pelagic species caught in large quantities specifically those of fatty meat such as tuna, mackerel and salmon (The AOCs Lipid Library 2011).

All three categories mentioned above are of scant edible use or the raw material is waste and is of no edible value and in some cases, specifically the latter two categories, they present a disposal problem.

Fish by-products

Fish waste and by-products are produced in a significant amount and are a key and direct consequence of fishing and fish processing. The amount of waste and by-products from fishing cause economical and environmental issues, as in some cases a proportion of the fish caught at sea are thrown back into the sea as waste. There are three main categories used to classify fish waste, these are discards, wastage on board and by-products and wastage on shore. Discards are defined as the portion of fish, dead or alive, that is thrown away or dumped at sea, although post-harvest waste such as offal are not taken into account (Kelleher 2005). By-catch, which is the total catch of non-target animals, is sometimes associated with discards. Many vessels process the capture on board, which produces sub-products such as heads viscera and skins. The sub-products are required to be frozen instantly to prevent them from perishing. Wastage on-board refers to these sub-products that perish on the vessels or discarded due to the low commercial viability on land. Finally, by-products and wastage on land refers to the by-products obtained from fish processing operations. The by-products include heads, viscera, tails, shells and fins that have a low acceptance by consumers or sanitary regulations that avoid their use in human foods (Berge and Galvez 2008). It is estimated that approximately 7.3 million tonnes of fish are discarded yearly (Kelleher 2005).

Salmon waste

Salmon is one of the key species of fish desired for human consumption. It is processed entirely for human consumption hence very high standards of freshness and hygienic quality are followed. This provides an advantage for the by-products manufacturers to produce high quality by-products (Wright 2004). It is estimated that salmon farming yields approximately 1 million tonnes of fish (Wright 2004) and according to FAO statistics the global fisheries produced roughly 2.5 million tonnes of salmon in 2010 (FAO 2012). From salmon farming it is estimated that 250,000 tonnes of salmon waste is available yearly to convert to useful products (Wright 2004). Salmon by-products can be classified through the anatomy of the fish, which include the head, viscera, bones, skin, tails and fin. An assessment of salmon by-products from the major salmon producers in 2004 was carried out by the FAO. The results can be seen in Table 3 (Ramirez 2007).

The oil content in salmon by-products can be found in the head, viscera, skin, and bones and is particularly high in the head and viscera (Ramirez 2007). For Atlantic salmon, every metric tonne of the by-products yields approximately 200 kg of orange coloured salmon oil and condensed salmon hydrolysate (Wright 2004). It has also been found that in pink salmon by-products, 10.9% oil can be found in the head and 2% in the viscera (Bechtel and Wu 2008).

Table 3 Salmon By-products (Adapted from Ramirez 2007).

Country	Production ('000tonnes/yr)	Heads, Bones & Skin ('000tonnes/yr)	Viscera ('000tonnes/yr)	By-product ('000tonnes/yr)
Chile	569	134	51	185
Norway	602	43	73	116
UK	141	16	9	25
Total	1312	193	133	326

Fish oil production

Raw fish is composed of three major fractions: solids (fat-free dry matter), oil and water. The aim of the manufacturing process is to separate these three major fractions as completely as possible with minimal expenditure and under optimum conditions to produce the best possible products (FAO 1986). There are various methods that can be used to reduce fish to fish oil and fishmeal, the key methods being wet pressing method, hydrolysis, silage production, dry rendering and solvent extraction. The principles of these methods are the same and the steps involved are common to all methods of practical importance. The general steps involve heating the fish to coagulate the protein and rupture the fat depots to liberate oil and physico-chemically bound water, pressing to remove a large fraction of the liquids from the mass, separation of the liquid into oil and water (stickwater), refining of the oil obtained by degumming, neutralization, bleaching and deodorisation stages (FAO 1986).

Chemical composition of fish oil

Fish oil is composed mostly of triglycerides of polyunsaturated fatty acids (PUFAs), monounsaturated fatty acids (MUFAs) and some saturated fatty acids (SFAs). In lesser and varying amounts can also be found: phospholipids, glycerol ethers and esters of waxes (FAO 1986). In addition 1–2% of fish oil consists of compounds such as squalene glycerides, carotenoids, sterols, tocopherols, chlorophylls, vitamin A and vitamin D (FAO 1986).

There are more than 60 different types of fatty acids that have been identified in fish oil. More than 80–85% of them are represented by four groups of fatty acids (Austic et al. 2010): (i) myristic (C14:0) and palmitic (C16:0), (ii) palmitoleic (C16:1n-7) and oleic (C18:1n-9), (iii) gadoleic (C20:1n-9) and erucic (C22:1n-9) and, (iv) eicosapentaenoic (C20:5n-3), docosapentaenoic (C22:5n-3) and docosahexaenoic (C22:6n-3).

Although there are two families of PUFAs (omega-6 and omega-3) present in fish oil, fish oil is rich in eicosapentaenoic acid (EPA) and docosahexaenoic acid (DHA) that are omega-3 fatty acids representing more than 90% of all PUFAs. Therefore, fish oil is more susceptible to oxidation (Austic et al. 2010). Omega-3 PUFAs are what differentiates fish oil from vegetable oil and animal oils. PUFAs are produced by phytoplankton and marine algae and are incorporated into the fish through the food chain (Shahidi 1998).

The main omega-3 PUFA contained in fish oil are alpha linolenic acid (C18:3n-3, ALA), eicosapentaenoic acid (C20:5n-3, EPA), docosahexaenoic acid (C22:6n-3, DHA) and in a lesser extent docosapentaenoic acid (C22:5n-3, DPA). The structure of these can be seen in Fig. 3.

The lipid content of fish oil varies depending on the species of fish, season and fishing area. Cod, tuna, anchovies and sardines are found to have a higher PUFA content. Salmon oil has been found to have the lowest PUFA content when compared to Atlantic shad, sardine, anchovy, cod and tuna (Shahidi 2006). Therefore, the waste by-products from salmon are more suited to provide fish oil to be converted to biodiesel. This is mainly due to the value of fish oil rich in PUFAs and the benefits to human health and aquaculture. Hence, it would not be deemed worthwhile to convert fish oil rich in PUFAs to biodiesel.

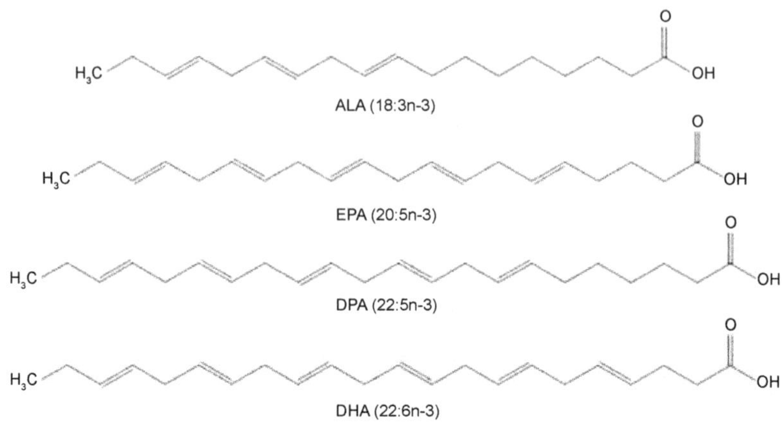

Fig. 3 Omega-3 Fatty Acids.

The price of fish oil

The price of fish oil has been generally increasing globally and reached a record high in March 2008, US $ 1700 per tonne. When the price of fish oil almost doubled from the previous year, March 2007, when the price was US $ 894 (Metian and Tacon 2008). The reason for the price increase can be attributed to many factors but in particular the fairly static global supplies of fish oil and fishmeal and the strong market demand for them by the aquaculture and livestock sector in the major importing countries. The price increase is also a consequence of the increasing key vegetable oil, global petroleum and energy prices. The demand for fish oil in aquaculture is continually growing, combined with a strong market for food supplements indicate high prices will be continued to be expected for fish oil in the near future.

Objectives

There is a current need to find a cheaper feedstock for the production of biodiesel, due to the problems associated with using food crops as feedstock. It is known that the fish processing industry produces large quantities of fish waste and by-products. As a consequence, there is a necessity to find a solution to this problem, which will enable an effective utilization of fish waste. One possibility is to use fish oil as fuel. However, using fish oil directly in the diesel engines causes many problems due to its high viscosity. Therefore, fish oil must be transformed into biodiesel in order to improve its fuel properties. Fish oil provides further problems in its use as biodiesel; the high polyunsaturated fatty acid (PUFA) content of fish oil makes it more susceptible to oxidation and the free fatty acid content of fish oil can reduce the biodiesel yield and complicate the separation and purification process.

Fish oil presenting important content in PUFAs have a higher commercial demand and value, due to the benefits to human health that omega-3 PUFAs provide. Hence, salmon oil, which presents a lower commercial application in the food industry due its low content in Omega-3 PUFAs, is an appropriate feedstock to produce biodiesel with good fuel qualities, especially in terms of oxidation stability, when compared to other types of fish oil that are high in PUFAs.

Thus, the purpose of this research was to study further the characteristics of biodiesel produced from salmon oil such as acid value, viscosity, density, FAMEs content, oxidative stability and cold flow properties. In this sense, the influence of the operating conditions chosen (temperature, ratio methanol/oil and reaction time) on the final biodiesel was evaluated.

MATERIALS AND METHODS

Materials

The fish oil employed contained 90% salmon oil and the remaining 10% was a mixture of other fish oil in season. It was purchased from Industrias Afines, S.L (Vigo, Spain). The fish oil presented the following properties: acid value of 6.6 mg KOH/g oil, peroxide value of 6, a composition of 6% w in EPA and 8% w in DHA, a bright brownish-orange color and a characteristic fish odour. All the reagents used were of analytical grade.

Production Process

The two-step process was used. The first step consisted of acid-catalysed pre-treatment to esterify the FFA, reducing the acid value of the raw fish oil below 2 mg KOH/g oil. 200 grams of salmon oil were reacted with methanol in a molar ratio of methanol to oil of 6:1 and 1% wt of sulphuric acid (96% purity) was employed as a catalyst. The temperature was set at 60°C and the reaction was carried out during 60 min. To overcome the mass transfer limitations, a stirring speed of 300 rpm was used. After the completion of the reaction, the content of the reactor was transferred to a separating funnel to settle for one hour, this allows the separation of the esterified salmon oil. In the second step, the esterified fish oil was reacted with methanol employing the corresponding methanol to oil molar ratio and 1% wt of sodium hydroxide was employed as alkaline catalyst. The temperature and time of reaction were set according to the design of the experiments. A stirring speed of 300 rpm was employed. The reaction products were transferred to a funnel to settle for one hour, the glycerol was separated from the alkyl esters and then purified by washing and drying.

Free Fatty Acids Analysis

The free fatty acid (FFA) content of the oil samples was determined according to the European standard EN 14104. The method used is based on titration of the sample, diluted with ethanol-diethyl ether mixed solvent, with potassium hydroxide solution in ethanol employing phenolphthalein as indicator to detect the end point. The results are expressed in milligrams of potassium hydroxide per gram of sample (mg KOH/g oil). The samples were measured in duplicate.

FAME Content Analysis

The salmon oil and biodiesel samples were both analyzed by gas chromatography. The salmon oil had to be first methylated by direct transesterification with acetyl chloride/methanol using the method provided by Lepage and Roy 1984 and modified by Rodríguez-Ruiz et al. 1998. Then the methyl esters of both the methylated salmon oil and biodiesel samples were analyzed according to the method described by Camacho et al. 2002 with a Agilent 7890A gas chromatograph (Agilent Technologies, S.A.) connected to a capillary column of fused silica Omegawax (0:25 mm × 30 m, 0:25 μm standard film; Supelco, Bellefonte, PA) and a flame-ionisation detector. Nitrogen was used as the carrier gas and the total column flow was 44 ml·min^{-1}. The oven temperature programme was initially set at 150°C for 3 min, then increased at a rate of 10°C·min^{-1} until 240°C and maintained at this temperature for 12 min. Matreya (Pleasant Gap, PA) n-3 PUFAs standard (catalogue number 1177) was used for the qualitative fatty acid determination for both the salmon oil and biodiesel. Nonadecanoic acid (C19:0) was used as an internal standard for quantitative determination of fatty acids for the salmon oil. Methyl heptadecanoate was used as the internal standard for the quantitative determination of fatty acids for the biodiesel samples.

Once the peaks of the biodiesel samples were identified, the peak areas were employed to determine the FAME content of each sample. Equation 1 was employed for this:

$$C = \frac{(\sum A) - A_{EI}}{A_{EI}} \times \frac{C_{EI-} V_{EI}}{m} \times 100 \qquad (1)$$

where C = fatty acid methyl ester content (%); ΣA = total peaks area; AEI = area of the peak corresponding to methyl heptadecanoate; CEI = concentration of methyl heptadecanoate solution ($mg \cdot ml^{-1}$); VEI = volume of methyl heptadecanoate solution (ml); m = mass of biodiesel sample (mg).

Viscosity Measurements

Viscosity of the biodiesel samples were measured using a rotational viscometer Haake model VT500 and NV sensor system (dual slot). Each sample was placed between the two coaxial cylinders and set at the desired temperature employing a water bath. Shear stress (τ) was measured at 30, 40 and 50°C at varying shear rates (γ) from 0 to 3000 s^{-1}. The dynamic viscosity (μ, Pa·s) of each sample was determined as the slope of the straight line resulting when plotting shear stress (τ, Pa) versus the shear rate (γ, s^{-1}) (Fig. 4).

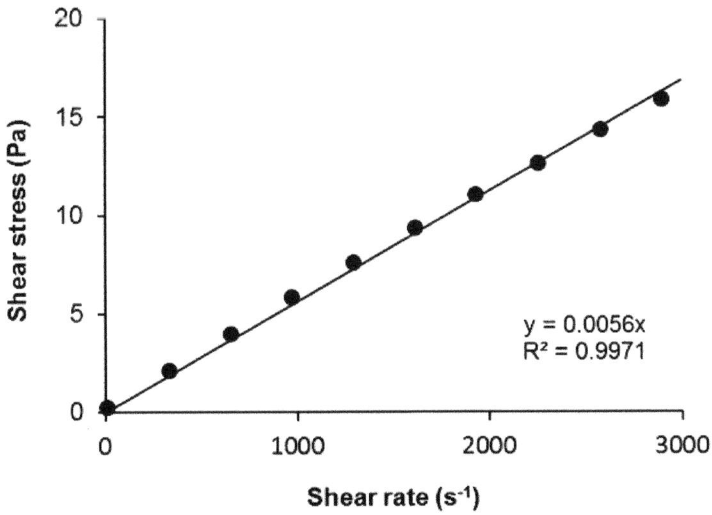

Fig. 4 Shear rate vs. Shear Stress Graph to Determine Dynamic Viscosity.

The effect of the temperature on viscosity was described through the Andrade relationship, shown in Eq. 2:

$$\mu = A\,exp\left(\frac{S}{T}\right) \qquad (2)$$

where, μ = dynamic viscosity (Pa·s); T = temperature (K), A = constant of the fluid (Pa·s) and B = constant of the fluid (adimensional).

Density Measurements

The density measurements involve heating the biodiesel sample to the required temperature and then simply weighing the sample, at that specific temperature, to determine the mass of the sample with a known volume. The measurements were carried out in duplicate.

Oxidative Stability Analysis

The Rancimat test was employed to determine the oxidative stability of the biodiesel samples according to EN 14112. A Metrohm Rancimat model 743 (Methrom Instruments, Herisau, Switzerland) was utilized. A stream of filtered, cleaned and dried air at a rate of 10 $L \cdot hr^{-1}$ was bubbled into 7.5 grams of biodiesel samples contained in reaction vessels. These vessels were placed in an electric heating block, which was set at the desired temperature. Effluent air containing volatile organic acids from the oil samples were collected in a measuring vessel with 60 mL of distilled water. The conductivity of the water was continuously recorded and the induction period (IP) was automatically determined using the apparatus. The measurements were carried out in duplicate at three different temperatures (90–100–110°C). Eight oil samples were analysed in the equipment at the same time. The rancimat induction period was expressed as the resistance time (in hr) of the oil to oxidation.

Differential Scanning Calorimetry Analysis

The DSC analyses of the biodiesel samples were conducted using a differential scanning calorimeter (Metler Toledo DSC 1, Metler Toledo). For each scan, approximately 10 mg of sample were hermetically sealed in an aluminium pan and tested against an identical empty pan. To determine the phase transition of the biodiesel samples, a ramp rate of $5°C \cdot min^{-1}$ was used under nitrogen atmosphere (50 $ml \cdot min^{-1}$). For heating scans, samples were cooled and held isothermally at –80°C for 10 min, then heated to 100°C. After the experiment was complete for each biodiesel sample, curves were obtained from the heating DSC scans and analysed to determine the following parameters: completion of melt onset temperature (COM), minimal heat-flow temperature of the high-melting peak (P1), and melting point (MP).

Experimental Design

The three experimental factors considered for this study were temperature, methanol to oil ratio and reaction time. A factorial experimental design comprising 27 runs was executed, in which each input variable was set at three levels: 40, 50 and 60°C for temperature; 3:1, 6:1 and 9:1 for methanol to oil ratio and 30, 60 and 90 min for reaction time.

RESULTS AND DISCUSSION

Acid value

The acid value is a measure of acidic substance in the oil and therefore the corrosiveness and long-term stability of the biodiesel. It can be used as a guide for quality control as well as for monitoring oil degradation during storage (Austic et al. 2010). The acid value was determined after the acid catalyzed reaction and also after the alkaline catalyzed reaction for all 27 experiments. The results have been provided in Table 4.

Table 4 Acid value.

Exp.	Temperature (°C)	Methanol:Oil Ratio	Time (min)	1st stage Acid Value (mg KOH/g Oil)	Final Acid Value (mg KOH/g Oil)
1	40	3	30	0.756	0.084
2	40	3	60	0.589	0.084
3	40	3	90	1.09	0.14
4	40	6	30	0.786	0.084
5	40	6	60	0.842	0.14
6	40	6	90	0.701	0.196
7	40	9	30	0.786	0.224
8	40	9	60	0.786	0.196
9	40	9	90	0.477	0.112
10	50	3	30	0.926	0.112
11	50	3	60	2.08	0.309
12	50	3	90	0.729	0.168
13	50	6	30	1.32	0.281
14	50	6	60	1.01	0.224
15	50	6	90	0.729	0.224
16	50	9	30	0.982	0.168
17	50	9	60	0.729	0.168
18	50	9	90	3.79	0.112
19	60	3	30	0.701	0.168
20	60	3	60	0.729	0.196
21	60	3	90	0.701	0.14
22	60	6	30	0.729	0.224
23	60	6	60	1.01	0.196
24	60	6	90	0.757	0.252
25	60	9	30	1.01	0.224
26	60	9	60	0.814	0.084
27	60	9	90	0.729	0.168

The purpose of the first stage of the two-step transesterification method was to reduce the free fatty acid content of the salmon oil by esterifying the free fatty acids to less than 1% FFA corresponding to 2 mg KOH/g oil (Canakei and Van Gerpen 2001). This was necessary to be able to carry out the base catalyzed transesterification reaction to completion. In previous research it has been found that the high FFA content of fish oil reduces the esters yield produced from using only the alkaline-catalyzed transesterification process and the separation of the esters becomes difficult (Avena-Bustillos et al. 2007). From Table 4, it can be seen that after the first step the acid value of the salmon oil has been reduced to less than 2 mg KOH/g oil for all experiments except experiments 11 and 18. For these two experiments the acid value is 2.08 mg KOH/g oil and 3.79 mg KOH/g oil respectively, although a successful alkaline transesterification was still achieved for these two experiments. Therefore, indicating that 3.79 mg KOH/g oil, which corresponds to 1.9% FFA, can be used as an upper limit for the acid value, to achieve an alkaline successful transesterification. Although this contradicts the results of Canakei and Van Gerpen 2001, it confirms the results of Avena-Bustillos et al. 2007. From the results, it has been found that a 6:1 methanol to oil ratio is more than sufficient to reduce the acid value from 6.6 mg KOH/g oil to generally less than 1.4 mg KOH/g oil.

The maximum acid number allowed according to the European standard is 0.5 mg KOH/g oil. From Table 4 it can be seen that after the alkaline-catalyzed transesterification of the triglycerides to complete the reaction, the acid value has been reduced to less 0.5 mg KOH/g oil for all experiments and thus meeting the European Standard. It can also be seen from Table 4 that the final acid value for biodiesel produced at all of the operating conditions studied is less than 0.31 mg KOH/g oil. Therefore, it was proved that the acid value can be reduced, after the second step, to meet the European standard at the minimum conditions assayed of 40°C, 3:1 methanol to oil ratio and a reaction time of 30 min.

Fatty Acid Methyl Ester Content (FAME)

A high content of FAME in the final biodiesel is required in order to present an acceptable fuel quality minimizing contaminants such as glycerine, methanol, mono-, di-, and triglycerides. These impurities can cause severe problems, such as engine deposits, filter clogging or fuel deterioration (Knothe et al. 2005b). Table 5 provides the fatty acid methyl ester (FAME) content for all the biodiesel samples.

From Table 5 it can be seen that at all of the conditions that were studied, the FAME content is greater than 93% for all samples. Therefore, a good conversion of triglycerides to FAMEs was obtained. The maximum

Table 5 FAMEs, PUFAs and SFAs Content.

Exp.	Temperature (°C)	Methanol:Oil Ratio	Time (min)	FAME (%)
1	40	3	30	93.12
2	40	3	60	93.26
3	40	3	90	93.72
4	40	6	30	93.98
5	40	6	60	93.65
6	40	6	90	93.76
7	40	9	30	93.58
8	40	9	60	93.35
9	40	9	90	93.53
10	50	3	30	93.08
11	50	3	60	93.27
12	50	3	90	93.43
13	50	6	30	93.87
14	50	6	60	93.91
15	50	6	90	94.19
16	50	9	30	93.6
17	50	9	60	93.59
18	50	9	90	95.14
19	60	3	30	94.19
20	60	3	60	93.95
21	60	3	90	94.65
22	60	6	30	94.05
23	60	6	60	94.1
24	60	6	90	93.56
25	60	9	30	93.15
26	60	9	60	94.78
27	60	9	90	95.39

conversion, producing 95.39% FAMEs, was achieved at 60°C with a 9:1 methanol to oil ratio and a reaction time of 90 min. Previous researchers have found temperature increase to have a positive effect on the FAME content up to 60°C and hence this result verifies those findings. This is due to the rate of reaction increasing as the temperature increases. Although it has also been found that methanol to oil ratio is generally more significant than reaction time (Jayaraj et al. 2004), the results of this research show that increasing reaction time has a slight increase in FAME content. According to Jayaraj et al. 2004, a lower methanol to oil ratio requires a longer reaction time. However, in the results depicted in Table 5 the methanol to oil ratio parameter seemed not to be important for the final FAME content. This can be attributed to the fact that the minimum methanol to oil ratio used was

3:1 at a minimum reaction time of 30 min and according to stoichiometry, a minimum of 3:1 is required for the reaction. Also, several researchers have stated that 30 min is sufficient for the reaction to reach completion (Ballesteros et al. 2007). Hence it may not have been possible to see the effect of the methanol to oil ratio fully due to the large reaction times used. In comparison the lowest percentage of FAME (93.12%) can be found at 40°C, with a 3:1 methanol to oil ratio and a reaction time of 30 min. This fact again reiterates the results found in previous research.

Final Biodiesel and Glycerine Mass and Yield

Once the two-step transesterification reaction was completed, a final weight of biodiesel was obtained after each experiment. The percentage yield was then calculated using the following equation (Eq. 3):

$$\%yield = \frac{mf \times \text{FAME}}{m_i} \tag{3}$$

where, m_f = final biodiesel mass (g); m_i = initial salmon oil mass (*g*); and FAME: percentage of FAME in the biodiesel sample. The final biodiesel mass, biodiesel percentage yield and the glycerine mass obtained from each experiment are displayed in Table 6.

From the experimental results obtained it was observed that experiment 7 produced the largest mass of biodiesel, 177g, and the highest percentage yield, 83.1%. The operating conditions for this experiment were 40°C, 9:1 methanol to oil ratio and a reaction time of 30 min. It was revealed that as the methanol to mole ratio increased, the final mass of biodiesel produced was considerably larger. Hence, this research confirms previous findings by Avena-Bustillos et al. 2007 and Jayaraj et al. 2004, as the largest biodiesel mass is produced at the highest methanol to oil ratio. The previous researchers also found that after 9:1 mole ratio there is no significant difference in the biodiesel yield produced. This cannot be confirmed through this research, as the maximum mole ratio used was 9:1. The optimum temperature for the maximum yield produced by Jayaraj et al. 2004 was found at 45°C ± 5°C and that the yield decreases at a reaction temperature greater than 50°C due to the saponification of glycerides by the alkaline catalyst before the alcoholysis is completed. The results found here agree with the findings of Jayaraj et al. 2004 and the decrease of biodiesel yield with increasing temperature can also be attributed to the difficulty of the separation process at higher temperatures. An increasing reaction time was found to have a positive influence on the biodiesel mass and yield found, as it allows a longer time for the completion of transesterification of triglycerides (Jayaraj et al. 2004).

Table 6 Final Biodiesel and Glycerine Mass and Yield.

Exp.	Temperature (°C)	Methanol:Oil Ratio	Time (min)	Final Biodiesel Mass (g)	Yield (%)	Glycerine Mass (g)
1	40	3	30	160.0	74.5	33.3
2	40	3	60	163.3	76.1	38.0
3	40	3	90	154.0	72.2	46.1
4	40	6	30	168.0	78.9	49.2
5	40	6	60	171.2	80.2	52.3
6	40	6	90	170.7	80.0	48.6
7	40	9	30	177.5	83.1	47.6
8	40	9	60	166.5	77.7	63.2
9	40	9	90	176.2	82.4	64.2
10	50	3	30	154.0	71.7	50.1
11	50	3	60	155.4	72.5	47.8
12	50	3	90	163.1	76.2	43.4
13	50	6	30	163.0	76.5	53.5
14	50	6	60	156.0	73.2	51.8
15	50	6	90	171.5	80.8	52.0
16	50	9	30	159.2	74.5	68.0
17	50	9	60	167.2	78.2	67.9
18	50	9	90	170.3	81.0	66.3
19	60	3	30	168.4	79.3	47.6
20	60	3	60	162.7	76.4	50.9
21	60	3	90	152.8	72.3	53.0
22	60	6	30	165.5	77.8	55.6
23	60	6	60	151.3	71.2	67.8
24	60	6	90	156.0	73.0	54.6
25	60	9	30	171.6	79.9	69.3
26	60	9	60	168.7	79.9	64.6
27	60	9	90	157.7	75.2	70.3

Further, from Table 6 it is observed that the maximum amount of glycerine, 70.3 g, is produced at 60°C, 9:1 mole ratio and a reaction time of 90 min. Also the minimum mass of glycerine, 33.3 g, is produced at 40°C, 3:1 mole ratio and 30 min. It was clearly seen that as temperature, methanol to oil mole ratio and reaction time increase so does the mass of glycerine produced. As glycerine is an unwanted by-product it is aimed to reduce the glycerine mass and increase the biodiesel yield produced. However, it was found that as the mole ratio and reaction time increases so does the FAME content and the weight of the biodiesel. The reaction temperature causes a dilemma for the production of biodiesel, as at the highest temperature, 60°C, produces the highest FAME content, 95.39%, but the lowest temperature,

40°C, provides the maximum mass, 177 g, and percentage yield, 83.1%, of biodiesel and minimum mass of glycerine, 33.3 g. Therefore, a compromise needs to be made for the temperature to maximize the FAME content and biodiesel yield whilst reducing the quantity of glycerine produced. Although using a lower weight percentage of the alkaline catalyst would allow a higher temperature to be used without minimising the biodiesel yield and improving the FAME content (Jayaraj et al. 2004).

Viscosity

The viscosity of a fuel is of major importance, as it controls the characteristics of the injection from the diesel injector. The viscosity in this case affects the atomization of a fuel upon injection into the combustion chamber (Knothe et al. 2005c). Hence, the high viscosity of a fuel is a major cause of poor fuel atomization, as the viscosity of the fuel increases the fuel will have a greater tendency to cause operational problems such as engine deposits on various engine parts (Knothe et al. 2005c). The viscosity of fatty acid methyl esters can reach very high levels; hence, it is important to control it within an acceptable level to avoid negative impacts on fuel injector system performance. Although there are engines that can use untransesterifed oils, the majority of engines require a lower viscosity fuel. Therefore, one of the foremost reasons that vegetable oils and animal fats are transesterified to alkyl esters is that the viscosity of biodiesel is significantly closer to the viscosity of petrodiesel (Knothe et al. 2005c). The viscosity of raw salmon oil was found to be 0.039 Pa·s at 30°C, 0.028 Pa·s at 40°C and 0.020 Pa·s at 50°C. In Table 7 it is shown that each biodiesel sample had a viscosity value approximately one order of magnitude lower than raw salmon oil. This proves the transesterification process is effective at reducing the viscosity of salmon oil. All samples showed Newtonian behavior over the experimental temperature range at which the viscosity was studied. The findings are also consistent with previous research (Avena-Bustillos et al. 2008).

Generally from Table 7 it is observed that for each temperature at which the viscosity was studied, the dynamic viscosity of the biodiesel produced for all 27 experiments does not change significantly at that specific temperature. For example, at 40°C the dynamic viscosity for all experiments is approximately 0.0045 Pa.s. This result is consistent with the study carried out by Avena-Bustillos et al. 2008 where a viscosity value of 0.0044 Pa.s was obtained for salmon oil methyl esters. In a previous study by Jayaraj et al. 2004, it was found that as the methanol to oil ratio increases, the viscosity decreases. In this research it was not clearly observed.

It is also shown that with increasing temperature the viscosity decreases. From the Andrade equation the minimum and maximum values found for the constants A and B were 0.00001 and 0.00005 Pa.s and 1426.3 and

Table 7 Dynamic Viscosity and Parameters of Andrade Equation.

Exp.	T (°C)	Methanol:Oil Ratio	Time (min)	Viscosity, mPa·s			Andrade's equation		
				30°C	40°C	50°C	A, mPa·s	B,K	r^2
1	40	3	30	5.62	3.97	3.86	0.0121	1846.5	0.816
2	40	3	60	5.42	4.47	3.89	0.0253	1624.2	0.994
3	40	3	90	5.42	4.49	3.85	0.0215	1674.6	0.999
4	40	6	30	5.45	4.67	4.00	0.0371	1511.8	1
5	40	6	60	5.37	4.46	4.00	0.0444	1449.7	0.983
6	40	6	90	5.50	4.53	3.79	0.0131	1830.7	1
7	40	9	30	5.39	4.50	3.85	0.0228	1655.4	0.999
8	40	9	60	5.33	4.44	3.80	0.0223	1658.9	0.999
9	40	9	90	5.49	4.60	3.85	0.0178	1737.8	1
10	50	3	30	5.55	4.57	3.95	0.0229	1662	0.996
11	50	3	60	5.46	4.46	3.81	0.0163	1760.1	0.997
12	50	3	90	5.50	4.51	3.86	0.0175	1740.6	0.997
13	50	6	30	5.32	4.44	3.77	0.0203	1687.7	1
14	50	6	60	5.41	4.43	3.82	0.0193	1705.8	0.995
15	50	6	90	5.40	4.50	3.82	0.0197	1700.1	1
16	50	9	30	5.30	4.43	3.91	0.0385	1490	0.992
17	50	9	60	5.36	4.69	4.00	0.0487	1426.3	0.996
18	50	9	90	5.44	4.50	3.95	0.0310	1563.1	0.993
19	60	3	30	5.42	4.61	3.91	0.0275	1601.7	0.999
20	60	3	60	5.35	4.57	3.96	0.0422	1467.1	1
21	60	3	90	5.41	4.54	3.81	0.0188	1716.2	1
22	60	6	30	5.37	4.39	3.90	0.0303	1564.8	0.983
23	60	6	60	5.08	4.40	3.75	0.0384	1481.4	0.998
24	60	6	90	5.47	4.47	3.73	0.0110	1881.8	1
25	60	9	30	5.34	4.43	3.87	0.0291	1576.8	0.995
26	60	9	60	5.40	4.40	3.67	0.0109	1880.1	1
27	60	9	90	5.51	4.64	3.89	0.0198	1707.1	0.999

1881.8 K, respectively. The coefficients of determination (R^2) obtained, values generally higher than 0.99, indicated that the change in the dynamic viscosity with temperature is well described by the Andrade equation.

In comparison with biodiesel from vegetable oils, the viscosity of biodiesel from fish oil is lower. This is due to the higher concentration of polyunsaturated fatty acids in biodiesel produced from fish oil.

Density

The density of fuel determines its energy per unit volume. This is of particular importance for a diesel engine as fuel is metered to the engine volumetrically. A fuel with low energy content per litre will cause the engine to produce less peak power (Knothe et al. 2005c). At part load conditions the engine operator will be able to meet the demand for power but a larger volume of fuel must be injected. In this case the fuel injection system may advance the fuel injection timing when the fuel flow rate increases and hence, cause an increase in nitrogen oxides emissions (Knothe et al. 2005c). Biodiesel has lower energy content on a weight basis as it contains methyl esters and not aromatics. However, methyl esters are denser and therefore, have more energy per unit volume.

Table 8 provides the density at 30°C, 40°C and 50°C for each batch of biodiesel produced at the various transesterification reaction conditions. The density values obtained versus temperature were fitted to linear models. The coefficients of the model as well as the determination coefficient (r^2) are also shown in Table 8.

Table 8 Density and Evaluation Parameters.

Exp.	T (°C)	Methanol:Oil Ratio	Time (min)	Density (kgm^{-3})			Linear model		
				30°C	40°C	50°C	Slope	Intercept	r^2
1	40	3	30	843.8	837.0	818.9	–1.25	883.1	0.937
2	40	3	60	839.8	831.1	819.7	–1.00	870.3	0.994
3	40	3	90	845.2	834.8	821.3	–1.19	881.5	0.995
4	40	6	30	840.3	820.1	808.3	–1.60	886.9	0.978
5	40	6	60	833.7	828.0	816.0	–0.88	861.3	0.959
6	40	6	90	837.4	828.5	824.0	–0.67	856.7	0.965
7	40	9	30	828.8	828.1	822.9	–0.30	838.5	0.843
8	40	9	60	838.3	834.2	823.5	–0.74	861.5	0.938
9	40	9	90	837.1	826.6	817.1	–1.00	866.8	0.999
10	50	3	30	839.4	829.5	827.0	–0.62	856.8	0.894
11	50	3	60	855.8	834.7	817.9	–1.90	911.9	0.996
12	50	3	90	851.9	831.1	826.1	–1.29	887.9	0.889
13	50	6	30	841.4	839.6	834.3	–0.36	852.6	0.925
14	50	6	60	845.4	838.3	834.0	–0.57	862.1	0.980
15	50	6	90	841.1	830.0	820.0	–1.05	872.5	0.999
16	50	9	30	832.2	835.6	827.0	–0.26	842.1	0.363
17	50	9	60	833.4	832.4	831.0	–0.12	837.2	0.991
18	50	9	90	845.7	832.5	813.4	–1.62	895.2	0.989
19	60	3	30	851.3	840.0	821.3	–1.50	897.5	0.980
20	60	3	60	836.4	835.7	826.1	–0.52	853.4	0.801
21	60	3	90	846.1	827.8	816.6	–1.47	889.0	0.981
22	60	6	30	832.1	826.6	819.4	–0.64	851.5	0.994
23	60	6	60	839.1	836.0	826.0	–0.65	859.8	0.914
24	60	6	90	841.1	833.5	819.4	–1.09	874.8	0.971
25	60	9	30	834.3	831.1	821.0	–0.67	855.5	0.919
26	60	9	60	831.9	827.3	821.9	–0.50	847	0.997
27	60	9	90	833.5	825.5	814.5	–0.95	862.6	0.992

It was revealed that as the measurement temperature increased density decreased for each sample. The degree of fit, shown by r^2 and being approximately 0.9 for each experiment, indicated that the changes of density with temperature were well described by the linear models proposed. Finally, it was generally observed that the methanol to oil molar ratio influenced on the final density. Higher values of density were obtained when employing lower methanol to oil molar ratio.

Oxidative Stability

Oxidation is a key problem for the storage of biodiesel. Biodiesel exhibits a relatively low level of storage stability as it is oxidized by the atmospheric oxygen. As fuel oxidizes, it may form acids and polymers that may cause fuel system deposits leading to fuel filter clogging and fuel system malfunctions. The fatty acid composition of the oil is one of the primary factors affecting the oxidative stability. The high number of double bonds present in the fatty acid chains increases the susceptibility to oxidation. The electronic representation of a double bond includes two components: a strong symmetry sigma bond and a pi bond. The pi bond is responsible for the greater reactivity of unsaturated compounds since it is weaker and has a lower energy (Austic et al. 2010).

The primary oxidation products being formed during oxidation are peroxides. After a while the esters are completely destroyed and the secondary oxidation products are formed. The secondary oxidation products formed include aldehydes, carboxylic acids, low molecular organic acids, other volatile organic compounds and substances of a polymeric nature, also referred to as sludge. The latter can cause motor damage. The oxidative stability of the fatty methyl esters is defined as its resistance to oxidation and it has key significance for quality control (Metrohm Ion Analysis 2010). Therefore, the oxidative stability of biodiesel is regularly determined during production and storage.

Table 9 provides the induction period determined by the Rancimat method at 90°C, 100°C and 110°C for each batch of biodiesel produced at the transesterification reaction conditions studied. The logarithms of the induction periods versus temperature were fitted to linear models and the gradients, intercepts and coefficients of determination for the different models obtained are also shown in Table 9.

It was observed that the temperature of measurement has a major influence on the determination of the induction period for the biodiesel samples produced. As expected, the IP values significantly decreased with temperature owing to the acceleration produced in the decomposition of lipid hydroperoxides (Frankel 2005) The degree of fit, as shown by the R^2

Table 9 Induction Period and Evaluation Parameters.

Exp.	T (°C)	Methanol:Oil Ratio	Time (min)	Induction period (h)			Linear model		
				90°C	100°C	110°C	Slope	Intercept	r^2
1	40	3	30	5.11	2.92	1.29	–0.0688	7.87	0.988
2	40	3	60	5.77	2.83	1.43	–0.0697	8.02	1.000
3	40	3	90	5.85	2.69	1.29	–0.0757	8.58	1.000
4	40	6	30	4.74	2.56	1.21	–0.0683	7.72	0.997
5	40	6	60	5.16	2.57	1.30	–0.0691	7.86	1.000
6	40	6	90	4.27	2.38	1.27	–0.0608	6.93	1.000
7	40	9	30	4.19	2.35	1.10	–0.0668	7.47	0.994
8	40	9	60	5.76	2.76	1.31	–0.0740	8.41	1.000
9	40	9	90	5.43	2.53	1.36	–0.0694	7.92	0.997
10	50	3	30	6.11	1.72	1.80	–0.0611	7.09	0.722
11	50	3	60	4.95	2.26	1.05	–0.0775	8.57	1.000
12	50	3	90	5.47	2.72	1.25	–0.0738	8.36	0.999
13	50	6	30	4.78	2.64	1.40	–0.0613	7.09	1.000
14	50	6	60	4.89	2.29	1.21	–0.0698	7.85	0.998
15	50	6	90	6.12	3.12	1.38	–0.0744	8.53	0.997
16	50	9	30	6.25	3.13	1.53	–0.0703	8.16	1.000
17	50	9	60	5.91	3.03	1.44	–0.0706	8.14	0.999
18	50	9	90	4.50	2.58	1.27	–0.0632	7.22	0.995
19	60	3	30	6.25	3.84	1.78	–0.0629	7.54	0.984
20	60	3	60	6.79	3.70	1.86	–0.0649	7.77	0.999
21	60	3	90	5.95	2.81	1.21	–0.0796	8.96	0.999
22	60	6	30	7.97	4.51	1.83	–0.0736	8.75	0.983
23	60	6	60	4.78	2.21	1.16	–0.0708	7.91	0.997
24	60	6	90	4.99	2.62	1.20	–0.0714	8.06	0.997
25	60	9	30	7.87	4.31	2.22	–0.0633	7.77	0.999
26	60	9	60	6.66	3.14	1.50	–0.0747	8.61	1.000
27	60	9	90	4.98	2.67	1.23	–0.0701	7.94	0.996

values of 0.72–1, indicates that the biodiesel oxidation rate in relation to temperature was well described by the linear models proposed of ln (IP) versus temperature.

From previous research it was found that at 110°C the oxidative stability for biodiesel produced from fish oil was only 0.4 hr (Austic et al. 2010). In contrast, from this study it was found that biodiesel produced from salmon oil exhibited an oxidative stability, at 110°C, that ranges from 1.05 to 2.2 h depending on the transesterification reaction conditions. Hence, there was a considerable improvement in the oxidative stability of the biodiesel produced. This may be attributed to salmon oil having a lower polyunsaturated fatty acid content than other types of fish oils such as tuna, anchovies and sardines.

However, the European standard for oxidative stability, given at 110°C, is a minimum of six hours. Therefore, the biodiesel produced from salmon oil in this study must be treated with antioxidants or blended with biodiesel produced from other types of oils that exhibit a higher oxidative stability (Austic et al. 2010) to improve its oxidative stability.

Cold Flow Properties

All diesel fuels are susceptible to start-up and performance problems when vehicles and fuel systems are subjected to cold temperatures. As ambient temperatures cool toward their saturation temperature, high molecular-weight paraffins (C18–C30 n-alkanes) present in petrodiesel begin to nucleate and form wax crystals suspended in a liquid phase composed of shorter-chain n-alkanes and aromatics. If the fuel is left under these conditions for an extended period of time, the solid wax crystals may cause start-up and performance problems. Despite the many advantages of biodiesel, as the transesterification reaction does not alter the fatty acid composition of the feedstock, it tends to have a poor performance in cold weather conditions. Biodiesel made from feedstocks containing higher concentrations of high-melting point saturated long-chain fatty acids tend to have relatively poor cold flow properties (Knothe et al. 2005c). Hence, this is one of the key disadvantages of biodiesel and must be resolved before biodiesel can become fully competitive with petrodiesel. Although, biodiesel produced from fish oils, due to its high polyunsaturated fatty acid content, present better cold flow properties than biodiesel produced from other feedstocks.

The cold flow properties most commonly used for biodiesel are cloud point, pour point, cold-filter plugging point and low-temperature flow test. Low temperatures cause the formation of solid wax crystal nuclei that are submicron in scale and invisible to the naked eye. If the temperature decreases further, these crystals grow. The temperature at which crystals become visible (diameter $\geq$ 0.5 μm), is defined as the cloud point as the crystals usually form a cloudy or hazy suspension (Knothe et al. 2005c). Hence, it is an index of the lowest temperature of the fuel's usability. When the engine is operated at a temperature below the cloud point, it can result in fuel filter clogging due to the wax crystals. At temperatures below the cloud point larger crystals (0.5–1 mm diameter × 0.01 mm thick) fuse together and form large agglomerates that can restrict or cut off flow through fuel lines and filters and cause start-up and performance problems. The temperature at which crystal agglomeration is extensive enough to prevent free pouring of fluid is determined by measurement of its pour point. The pour point is the temperature at which the fuel ceases to flow. Hence, the pour point indicates the lowest temperature of the fuel's utility (Knothe et al. 2005c). The cold-filter plugging point and low-temperature flow test are employed to determine the operability limits of the fuel, as operability temperature limits cannot be accurately predicted using the cloud point and pour point measurements (Dunn 1999).

According to Dunn 1999, it is possible to use DSC cooling or heating curves to obtain correlations with the cold flow properties of biodiesel. From the DSC thermograms, Fig. 5 is given as an example, it was possible to see three distinct endothermic peaks for all of the biodiesel samples and the melting point ranged from approximately between –73°C and 4°C for all samples. Sathivel 2005 found that for red and pink salmon oil the melting points were –69.6°C to –0.36°C and –64.7°C to 20.8°C, respectively. Similarly, Sathivel et al. 2008 reported a melting point range of –69.5°C to 14.2°C for unrefined Pollock oil. The negative melting point of the biodiesel samples can be attributed to the high content of unsaturated fats (Sathivel et al. 2008). This is due to oils with a higher unsaturated fatty acid content melting at negative temperatures, whereas those with higher saturated fatty acid content melt at a higher temperature. The first endothermic peak visible was not very sharp. This may be associated with the presence of impurities such as methanol and phospholipids. The next endothermic peak (P1) was

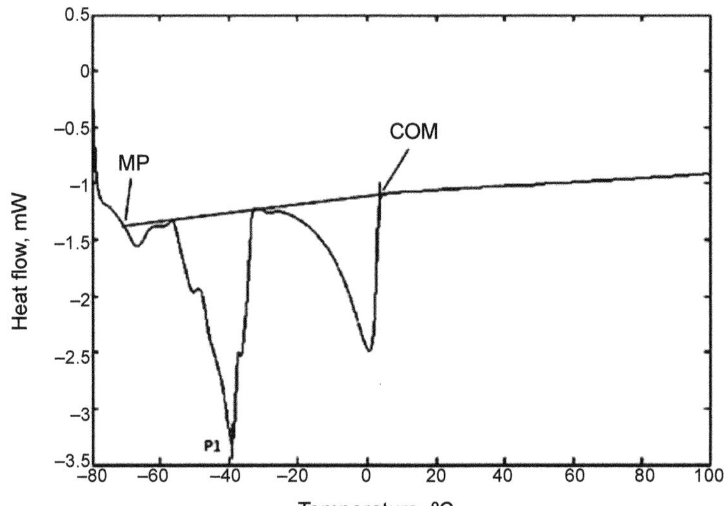

Fig. 5 DSC Heating Curve.

very distinct and sharp, at approximately –39°C. This peak corresponds to low-temperature freezing crystals, associated with methyl esters of unsaturated fatty acids. Finally, the last peak visible, seen at around 1–4°C, can be associated to the higher temperature freezing crystals, methyl esters of saturated fatty acids. Garcia-Perez et al. 2010 reported similar findings for biodiesel obtained from poultry fat. They reported the melting range to be between –30°C and 15°C. They also stated that the methyl esters of unsaturated fatty acids were associated to the lower temperature of –30°C and the methyl esters of saturated fatty acids were associated with the higher temperature of between –10°C and 15°C.

Completion of melt onset temperature (COM) was defined as the intersection of a line tangent to the scan at the point of the sharpest slope and a line tangent to a segment of baseline on the "hot side" (Dunn 1999). The COM temperature, of approximately 4°C for all samples is shown in Table 10. The melting point (MP) was defined as the point where a line

Table 10 DSC Results.

Exp.	Temperature (°C)	Methanol:Oil Ratio	Time (min)	COM (°C)	P1 (°C)	MP (°C)
1	40	3	30	3.70	–39.0	–73.3
2	40	3	60	3.71	–38.7	–62.5
3	40	3	90	3.72	–38.8	–73.2
4	40	6	30	3.70	–38.9	–73.6
5	40	6	60	3.75	–38.8	–73.6
6	40	6	90	3.83	–38.8	–72.2
7	40	9	30	3.76	–38.9	–73.3
8	40	9	60	3.79	–38.8	–72.3
9	40	9	90	3.82	–38.9	–73.1
10	50	3	30	3.70	–44.6	–68.4
11	50	3	60	3.73	–39.0	–72.2
12	50	3	90	3.55	–39.0	–73.1
13	50	6	30	3.69	–44.1	–69.2
14	50	6	60	3.75	–38.8	–72.1
15	50	6	90	3.48	–38.9	–72.3
16	50	9	30	3.73	–38.8	–72.7
17	50	9	60	3.82	–38.9	–64.8
18	50	9	90	3.58	–38.9	–72.1
19	60	3	30	3.72	–43.6	–66.1
20	60	3	60	3.74	–38.8	–73.0
21	60	3	90	3.44	–38.8	–70.4
22	60	6	30	3.54	–39.0	–72.6
23	60	6	60	3.74	–39.0	–72.5
24	60	6	90	3.37	–39.1	–72.7
25	60	9	30	3.31	–39.2	–73.0
26	60	9	60	3.69	–38.9	–71.8
27	60	9	90	3.77	–38.7	–70.5

tangent to the scan at the point of the first slope on the "cool side" of the melting peak intersects a line across the base of the peak (Dunn 1999). The MP temperature was found to be between –73°C and –62°C for all samples, depicted in Table 10. Also the minimal high melting peak temperatures (P1) for the 27 biodiesel samples are shown in Table 10.

CONCLUSIONS

This chapter investigated the feasibility of producing biodiesel from fish oil. It was determined that after the first stage of the process, the acid value was reduced to less than 2 mg KOH/g oil and after the second step of the two-stages transesterification process that at all of the reaction conditions studied it was possible to reduce the acid value to less than 0.31 mg KOH/g oil. It was found that for all experiments the FAME content of the biodiesel was greater than 93%. The FAME content of the biodiesel can be improved to beyond 96.5%, as given by the European standard, by using a lower alkaline catalyst quantity in the second step of the production process, such as 0.5% wt. Acceptable values of viscosity were obtained for the biodiesels produced. However, blending the biodiesel with petrodiesel may further reduce these values. The oxidative stability of all samples at 110°C ranged between 1 and 2 hour. It was revealed that salmon oil exhibited a longer oxidative stability than biodiesel produced from other types of fish oils. The oxidative stability can be increased to meet the European standard of six hours through the use of antioxidants. Finally, biodiesel from salmon oil provided good cold flow properties with a completion of melt onset temperature (COM) of approximately 4°C for all samples.

ACKNOWLEDGMENTS

This work was supported by the Spanish National Plan I+D+i (projects CTQ2008-02978 and CTQ2011-23009).

REFERENCES

Austic, G., R. Burton and X. Fan. 2010. Preparation and characterization of biodiesel produced from fish oil. Chem. Technol. Fuels Oils 46: 287–293.

Avena-Bustillos, R.J., P.J. Bechtel, B.S. Chiou, R.O. Dunn, H.M. El-Mashad, G.M. Glenn, S.H. Imam, T.H. Mchugh, W.J. Orts and R. Zhang. 2008. Biodiesel from waste salmon oil. Trans. ASAE 51: 797–802.

Avena-Bustillos, R.J., H.M. El-Mashad and R. Zhang. 2007. A two-step process for biodiesel production from salmon oil. Biosys. Eng. 99: 220–227.

Ballesteros E., M.P. Dorado, F.J. Lopez and M. Mittelbach. 2007. Optimization of alkali catalyzed transesterification of *Brassica Carinata* oil for biodiesel production. Energ. Fuel. 18: 77–83.

Bechtel, P.J. and T.H. Wu. 2008. Salmon by-product storage and oil extraction. Food Chem. 111: 868–871.
Berge, J.P. and R.P. Galvez. 2008. General introduction about by-products: worldwide situation and French focus. *In*: J.P. Bergé [ed.]. Added value to fisheries waste. Transworld Research Network, Kerala, India pp. 1–22.
Camacho Paez, B., A. Robles Medina, F. Camacho Rubio, P. González Moreno and E. Molina Grima. 2002. Equilibrium and kinetics: production of structured triglycerides rich in n-3 polyunsaturated fatty acids by the acidolysis of cod liver oil and caprylic acid in a packed-bed reactor. Chem. Eng. Sci. 57: 1237–1249.
Canakei, M. and J. Van Gerpen. 2001. Biodiesel production from oils and fats with high free fatty acids. Am. Soc. Agric. Eng. 44: 1429–1436.
Demirbas, A. and S. Karslioglu. 2007. Energy sources: biodiesel production facilities from vegetable oils and animal fats. Energ. Source, Part A 29: 133–141.
Dunn, R.O. 1999. Thermal analysis of alternative diesel fuels from vegetable oils. J. Am. Oil. Chem. Soc. 76: 109–115.
FAO. 1986. The Production of Fish Meal and Oil. Available at: http://www.fao.org.
FAO. 2012. Global Production Statistics 1950–2010: Diadromous Fishes/Salmon. Available at: http://www.fao.org.
Frankel, E.N. 2005. Lipid Oxidation (2th ed.). The Oily Press, Bridgwater, UK.
Guérard, F. 2006. Enzymatic Methods for Marine By-Products Recovery. *In*: F. Shahidi [ed.]. Maximising the value of marine by-products. Woodhead Publishing Limited, Cambridge, UK pp. 108–143.
Jayaraj, S., C. Muraleedharan and A.S. Ramadhas. 2004. Biodiesel Production from High FFA Rubber Seed Oil. Fuel. 84(4): 335–340.
Kelleher, K. 2005. Discards in the World's Marine Fisheries—An update. FAO Fisheries Technical Paper 470. FAO Fisheries Department.
Knothe, G. Introduction. 2005a. *In*: G. Knothe, J. Van Gerpen and J. Krahl [eds.]. The Biodiesel Handbook. AOCS Publising, Champaign, ILL, USA pp. 9–12.
Knothe, G. 2005b. Analytical methods for biodiesel. *In*: G. Knothe, J. Van Gerpen and J. Krahl [eds.]. The Biodiesel Handbook. AOCS Publising, Champaign, ILL, USA pp. 70–84.
Knothe, G. 2005c. Fuel properties. *In*: G. Knothe, J. Van Gerpen, and J. Krahl [eds.]. The Biodiesel Handbook. AOCS Publising, Champaign, ILL, USA pp. 84–163.
Lepage, G. and C. Roy. 1984. Improved recovery of fatty acid through direct transesterification without prior extraction or purification. J. Lipid Res. 25: 1391–1396.
Meher, L.C., D. Vidya Sagar and S.N. Naik. 2006. Technical aspects of biodiesel production by transesterification—A review. Renew. Sust. Energ. Rev. 10: 248–268.
Metrohm Ion Analysis. 2010. Determination of the Oxidative Stability of Biodiesel. Available at: http://products.metrohm.com
Metrohm Ion Analysis. 2010. Titrimetric analysis of biofuels. Available at: http://products.metrohm.com.
Ramirez, A. 2007. Salmon By-Products Proteins. FAO Fisheries Circular No. 1027. FAO Fisheries Department.
Rodríguez-Ruiz, J., E. Belarbi, J.L. García Sánchez and D. López Alonso. 1998. Rapid simultaneous lipid extraction and transesterification for fatty acids analysis. Biotechnol. Tech. 12: 689–691.
Tacon, A.G.C. and M. Metian. 2008. Global overview on the use of fish meal and fish oil in industrially compounded aquafeeds: Trends and future prospects. Aquaculture 285: 146–158.
Sathivel, S. 2005. Thermal and flow properties of oils from salmon head. J. Am. Oil Chem. Soc. 82: 147–151.
Sathivel, S., J. Huang and W. Prinyawiwatkul. 2008. Thermal properties and applications of the arrhenius equation for evaluating viscosity and oxidation rates of unrefined pollock oil. J. Food Eng. 84: 187–193.

Shahidi, F. 1998. Functional Seafood Lipids and Proteins. *In*: G. Mazza [ed.]. Functional Foods: Biochemical and Processing Aspects (Vol. 1). Technomic Publishing Company, New Holland, PA, USA pp. 381–402.

The AOCs Lipid Library. 2011. The Production and Processing of Marine Oils. Available at: http://lipidlibrary.aocs.org/processing/marine/index.htm.

Van Gerpen, J. and G. Knothe. 2005. Biodiesel production. *In*: G. Knothe, J. Van Gerpen and J. Krahl [eds.]. The Biodiesel Handbook. AOCS Publishing, Champaign, ILL, USA pp. 34–70.

Wright, I. 2004. Salmon By-Products. Aqua Feeds: Formulation & Beyond. 1: 10–12.

Index

For Product Safety Concerns and Information please contact our EU
representative GPSR@taylorandfrancis.com
Taylor & Francis Verlag GmbH, Kaufingerstraße 24, 80331 München, Germany

www.ingramcontent.com/pod-product-compliance
Ingram Content Group UK Ltd.
Pitfield, Milton Keynes, MK11 3LW, UK
UKHW020200200926
475702UK00004B/549

* 9 7 8 1 4 6 6 5 8 5 7 9 9 *